exoplanets

exoplanets

DIAMOND WORLDS, SUPER EARTHS, PULSAR PLANETS, AND THE NEW SEARCH FOR LIFE BEYOND OUR SOLAR SYSTEM

MICHAEL SUMMERS
JAMES TREFIL

SMITHSONIAN BOOKS
WASHINGTON, DC

The authors dedicate this book to our fellow Americans who are living with multiple sclerosis and Parkinson's disease.

Don't give up!

This book may be purchased for educational, business, or sales promotional use. For information, please write: Special Markets Department, Smithsonian Books, P.O. Box 37012, MRC 513, Washington, DC 20013

Published by Smithsonian Books
Director: Carolyn Gleason
Managing Editor: Christina Wiginton
Production Editor: Laura Harger
Edited by Emily Park
Designed by Jody Billert
Typeset by Scribe

Library of Congress Cataloging-in-Publication Data
Names: Summers, Michael E., author. | Trefil, James, 1938–
Title: Exoplanets : diamond worlds, super Earths, pulsar planets, and the new search for life beyond our solar system / Michael Summers, James Trefil.
Description: Washington, DC : Smithsonian Books, [2017] | Includes index.
Identifiers: LCCN 2016018596 | ISBN 9781588345943 (hardcover)
Subjects: LCSH: Extrasolar planets. | Planets. | Life on other planets. | Extraterrestrial beings.
Classification: LCC QB820 .S86 2017 | DDC 523.2/4—dc23 LC record available at https://lccn.loc.gov/2016018596

ISBN 978-1-58834-625-4 (paperback)

Manufactured in the United States of America
21 20 19 18   5 4 3 2 1

contents

1

NOT YOUR GRANDFATHER'S GALAXY

*There are more things in heaven
and earth, Horatio, than are
dreamt of in your philosophy.*

Shakespeare, Hamlet, *Act 1, Scene 5*

The universe used to be a simple place. We lived in a sedate solar system with nine planets circling an ordinary star in an unremarkable part of the galaxy. We assumed there were other solar systems out there—systems pretty much like ours. We had genteel arguments about whether some of those systems might support life, and we enjoyed science fiction adventure stories such as *Star Trek* and *Star Wars* that populated the galaxy with interesting (and often combative) beings who spoke English. But the central fact was that we knew about only one planetary system, so we labored under what we can call "the curse of the single example."

If you have only one example of something—be it a planetary system or a butterfly—the natural assumption is that every other thing you find will be like the one you know about. Take butterflies as an example. If the only kind of butterfly you had ever seen was a monarch, it would be reasonable to assume that all butterflies have to be big and orange and migrate to a particular spot in California every year. Confronted with a cabbage butterfly—small, white, and nonmigratory—you might understandably be confused. Some of your scientific colleagues might even argue that what you were seeing wasn't a butterfly at all, but a kind of beetle. Eventually, though, you would begin to explore a little more and find that the discovery of the cabbage butterfly was just the beginning of a journey into a world of amazing complexity and diversity, and that there were thousands of different kinds of butterflies in nature. You would realize that your original paradigm—the notion that there was only one kind of butterfly—was simply wrong and that it had blinded you to the true complexity of the living world.

We argue in this book that the butterfly analogy is a perfect description of humanity's recent discovery of the universe of exoplanets: planets outside our solar system. Only 30 years ago, most scientists would have asserted that we had a perfectly good explanation of the origins of our own solar system, an explanation based on the solid bedrock of the laws of physics and chemistry. These laws, they would have said, dictate that any other solar systems out there would have an inner contingent of small, rocky planets and an outer set of gas giants. These other solar systems, in other words, would be just like ours. And like the hypothetical butterfly collector in our analogy, we would begin our exploration of the worlds beyond our solar system with the wrong paradigm in mind, and, again like that hypothetical collector, we would be overcome by the incredible complexity we found when we actually looked at what is out there.

Planetary surprises were not slow in coming. Before we even got out of our own backyard, the way we looked at our solar system underwent a revolution. We began to see that, instead of a handful of planets in sedate orbit around the Sun, the moons of the outer planets constitute a group of diverse worlds in their own right. One of them, Jupiter's moon Europa, turned out to have a vast ocean of liquid water under its icy exterior, a fact that instantly made it a target for scientists interested in finding life away from Earth. Since that early discovery, such interior oceans have been found on other Jovian moons; on Enceladus, a moon of Saturn; and perhaps even under the frozen surface of Pluto. Instead of being a rarity found only on Earth, liquid water appears to exist in many other places even within our own solar system. The paradigm that told us that water has to be in surface oceans, as on Earth, was just wrong.

Things got more curious as we started exploring the outer reaches of our system. We'll touch briefly on the silliness involved in the "demotion" of Pluto in chapter 4, but the fact of the matter is that Pluto is actually the gateway to a whole new part of the solar system. Called the Kuiper belt after the Dutch astronomer Gerard Kuiper (1905–73), who suggested its existence in 1951, this is a flat disk of material that extends out beyond Pluto. We have known about the belt for a long time, but it was usually considered a kind of afterthought to the inner planets. Indeed, one of the authors of the book you are holding (James Trefil, hereafter JT) once compared it to a scrap pile left at a construction site after the important building was done.

This attitude changed quickly when astronomers discovered that, far from being an inconsequential pile of rubble, the Kuiper belt is actually home to an incredible variety of planets. Some of these planets are the size of Pluto, and some even have moons. Today, some astronomers estimate that dozens of planets may be lurking out there, a number that completely dwarfs the familiar inner group that includes Earth. Even before we left the solar system, in other words, the simple paradigm of "nine planets orbiting the Sun" was breaking down. Instead of being a lonely, demoted outsider, Pluto became the beginning of a previously unknown collection of worlds.

The Search for Exoplanets

Our search for planetary systems circling other stars has a long history. We'll discuss the daunting problems involved in this search in chapter 3. Even so, as you might guess, when we finally nailed down the existence of such a system in 1992, the discovery came as a complete surprise. The new planets, which are indubitably there, turned out to be circling the wrong kind of star, a kind of star called a pulsar. Pulsars are small, unbelievably dense, rapidly rotating masses of matter left when a large star explodes in a supernova. These supernova events mark the end of the line in the evolution of some types of stars. The titanic explosion blows huge amounts of material out into space, and you would expect that any planet unfortunate enough to be in orbit around such a star would be completely destroyed. Yet here these planets are, where no planet ought to be.

If the pulsar planets were the first surprise, the detection of planets circling normal stars was the next. The technique originally available for exoplanet detection, described in detail in chapter 3, involved measuring the small motion of the star ascribable to the gravitational pull of its planet. Such a technique is best at detecting large planets—those capable of exerting strong gravitational pulls on their star. Someone observing our own solar system with this technique, for example, would see the effects of Jupiter before he or she (or it) saw the effects of Earth.

In any case, when this technique was used to search for exoplanets, the first positive results were the discovery of what came to be called hot Jupiters. These are massive planets—typically several times larger than Jupiter—orbiting close to their stars, often closer to their stars than Mercury is to ours. But according to the paradigm that other solar systems should be like ours, this was impossible. Gas giants such as Jupiter were supposed to form only far away from their star, not close in. Another surprise; another failure of the paradigm. As the collection of hot Jupiters grew, astronomers began to wonder if *any* system out there is like ours.

As it turned out, they need not have worried. The fact that we were finding hot Jupiters first was simply a result of the detection system available. The situation changed with the launch of the Kepler satellite in 2009. We'll describe this incredible instrument in more detail in chapter 5, but basically it searches for the small dimming of a star's light due to the passage of a planet across the star's face.

It's important to realize that this type of search will be successful only if the orbit of the planet is oriented so that the planet passes between its star and Earth. A planet whose orbital plane is perpendicular to that line of sight is invisible. Also, the satellite searched only a small segment of the sky—think of it as searching an area a couple of times bigger than a full moon. Despite the limited nature of the search, however, Kepler found over 4,000 exoplanet systems in its four years of operation.

Talk about surprises! The first surprise that came from the Kepler satellite was the sheer number of exoplanets out there. Extrapolating from the small volume that Kepler searched to the entire galaxy, astronomers quickly realized that the Milky Way must contain more planets than stars. Far from being a rare event, in other words, the formation of planetary systems seems to be pretty much the norm. Like the butterfly collector in our example, we are having to adjust to the notion that the universe is a lot more complex and diverse than we imagined.

After that initial shock, surprises continued to emerge. As we refined our detection techniques, all sorts of new and strange worlds began to show up. Hot Jupiters faded into the background and a complex array of planets came into sight. These are discussed in detail in later chapters, but the new assortment of planets includes:

• Super Earths—rocky planets several times the size of the Earth. There seem to be a lot of these out there.
• Styrofoam worlds—planets so light that we cannot figure out why they don't collapse under their own gravity.

- Diamond planets—planets made of pure carbon, with diamond mantles and cores of liquid diamond, a material unknown on Earth.
- Multistar worlds—planets that circle up to four stars, systems that were supposed to be dynamically impossible.
- Hot Earths—worlds so close to their stars that their surface rocks are vaporized. When such a planet rotates, "snowflakes" made of solid rock fall from the sky.
- Rogue planets—planets wandering around unattached to stars. It is possible that the majority of planets in the galaxy are of this type.

Faced with this incredible (and growing) diversity, we have to give up our old ideas about how planetary systems form and recognize that our own system is only one of many types that can exist. We must, in other words, develop new paradigms to deal with what we are learning about exoplanets.

As this list of strange worlds grows, we have begun to realize that the intense concentration on what has come to be called the Goldilocks planet was simply misplaced. The Goldilocks planet is a hypothetical body that, like Earth, is situated in a position near its star that makes it "not too hot, not too cold, but just right." By "just right," we mean that it can have oceans of liquid water on its surface. The reason for this concentration, of course, is that we might expect such a world to be the home of life like ours.

What about Life?

And this brings us to the issue that generates the most interest in exoplanets: the question of whether any of these new worlds is a home to life. Once we turn our attention to the existence of life, however, we have to realize that we are once again confronting the curse of the single example. We know of only one type of life, the result of only one experiment. At the most basic molecular level, every living thing on Earth is descended from a single first cell and operates through the use of the same genetic code, the same basic DNA structure. At the molecular level, you have a lot more in common with the grass on the lawn than you might think. As we did when we first began exploring the realm of exoplanets, we approach the question of life with the assumption that whatever we find out there (if anything) will be "like us" to some degree.

We can think of the origination of life on Earth as occurring in two stages, rather like gears shifting in a car. The first stage was the development of the first living cell from inorganic materials, and the second was the process by which that first cell produced the diversity of living forms we see around us today.

We actually have a pretty good notion of how life evolved on Earth once the first cell showed up—it's contained in the theory of evolution. Some of the pieces of the puzzle involved in how that first cell developed are in place, and intense research efforts are being carried out to fill in the gaps. We know that life established itself on our planet 3.5 billion years ago, and that for the next 3 billion years Earth was a pretty dull place. An extraterrestrial visiting Earth then would have found a planet whose oceans were full of green pond scum. It is only in the past half billion years that complex multicelled life showed up, with intelligence and technology appearing much later than that. We can expect, then, that even if we do find life on an exoplanet, the discovery will most likely be that of a "pond scum planet."

The prevailing paradigm is that any life we find out there will be carbon based and will operate in a way similar to that of life on Earth, although not necessarily with the same molecules. If life is based on molecular chemistry, as it is on Earth, there will have to be some molecular mechanism that plays the same role as DNA in passing genetic information from one generation to the next. Such a molecule will have to be large and complex, and, so the argument goes, it will have to involve carbon chains. Carbon chemistry proceeds most quickly in liquid water, and this explains why we are searching for the Goldilocks planet.

Yet even if we confine our attention to molecular-based life, the sheer number and variety of exoplanets suggest that we should be prepared for surprises, for patterns that we don't see on Earth. To mention just one example, Earth's pattern of natural selection and evolution is driven in part by the fact that plate tectonics is constantly shifting the geography of the planet, constantly changing ecosystems. This means that organisms are constantly playing catch-up, constantly trying to adapt to new realities. It has been suggested, for example, that the development of upright posture and intelligence in early humans was driven by the drying up of rain forests in north-central Africa millions of years ago. We can ask, however, what evolution would look like on a world without a constantly changing surface. Would it come to a stop? Would the progression in complexity we see in Earth's fossil record show up on such a world? Would intelligence and technology evolve? Somewhere out in the array of exoplanets are the answers to questions such as these.

There are deeper questions we can ask, too: Does life really have to be based on molecular chemistry? Does it have to evolve according to the dictates of natural selection, as it does on Earth? It has become a standard quip among scientists that life is like pornography—we can't define it, but we know it when we see it. We argue that this may not be true and try to stretch our imaginations by suggesting the possibility of entities that are (arguably) alive but are not "like us." In chapter 12, we suggest that, just as we needed a new paradigm to deal with exoplanets, we will need a new paradigm to deal with life—a paradigm that inevitably takes us away from the Goldilocks planet and toward something much richer and more exciting.

The marvelous variety of planets actually raises an old question known as the Fermi paradox. Named after the Italian American physicist Enrico Fermi (1901–54), it involves an incident in which, after hearing an argument that the galaxy should be full of advanced technological civilizations, he asked a simple question: "Where is everybody?" Given the rich variety of worlds we know to be out there, why do we seem to be alone?

In the end, this might be the most important question raised by our new knowledge of the galaxy.

A Word about Chauvinisms

Nicolas Chauvin was a legendary character in French folklore—a man whose enthusiasm for his country was so great that his name is now attached to any attitude that involves (per the dictionary) "excessive or prejudicial support for one's own ideas."

In the sciences, the term *chauvinism* is usually used pejoratively to describe a position that precludes inquiry beyond generally accepted ideas. The ancient Greeks, for example, might be described as geocentric chauvinists, since they refused to consider the possibility that the Earth was not the center of the universe.

Thinking about exoplanets, and particularly about the nature of life on those planets, is full of chauvinisms. The most common of these—carbon chauvinism—is the notion that life elsewhere must be based on carbon. (For the record, both authors plead guilty to being carbon chauvinists, for reasons that will be explained later.) There are, however, two other chauvinisms that, while widespread, are less well known. We call them "surface chauvinism" and "stellar chauvinism."

Surface chauvinism is the idea that life has to exist on the surface of planets. This is what lies behind the search for the Goldilocks planet, for example. Yet it is actually a strange notion for a terrestrial scientist to have. After all, we know that on our planet there are complex ecosystems at deep-sea vents, miles below the ocean surface. We also know that there is bacterial life in rocks miles below the surface of the continents—in fact, some biologists have argued that such organisms make up the majority of Earth's biomass. Why should exoplanets be different?

Stellar chauvinism lies a little deeper in our subconscious. It is the idea that planets—at least planets capable of supporting life—have to be in orbit around stars. One of the great shocks in the search for exoplanets has been the dawning realization that most of the planets in our galaxy are *not* attached to stars. We call these "rogue planets" and imagine a visit to one in chapter 7. We discuss what kinds of life they might support in chapter 13.

We don't know what other kinds of chauvinisms may be hidden in our minds, but recognizing the ones outlined above is a good way to start our journey through exoplanets both real and imaginary.

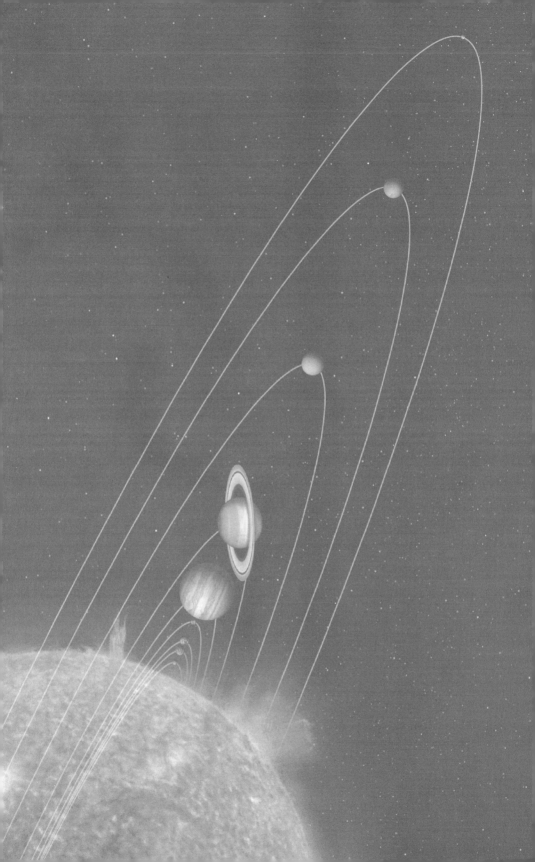

2

OUR BACKYARD

*The journey of a thousand miles
begins with a single step.*

Chinese proverb

t makes sense that the first region that humans would
explore when they turned their eyes to the heavens would be
our own solar system. After all, in astronomical terms, the
Moon and our sister planets are our neighbors, closer to us than
anything else in the sky. How could we avoid investigating them
first?

The origin of human knowledge of the solar system is lost
in the mists of antiquity. Monuments such as Stonehenge show
that people with no written language and, indeed, no metal tools
could nonetheless possess an astonishingly detailed knowledge
of the motion of the Sun, the Moon, and the planets. We know that
the great ancient civilizations were aware of the planets—the
Greeks, for example, assigned each of them to its own crystal

sphere in the sky, and the appearances and disappearances of Venus may have influenced the development of the Mayan number system. Nevertheless, it wasn't until 1609 that Galileo Galilei (1564–1642) turned a telescope to the sky and encountered the first of what would turn out to be a long string of paradigm-smashing surprises.

In Galileo's time, the cosmology that was taught in universities, and that Galileo presumably would have learned, was that of the ancient Greek philosophers. This cosmology taught that Earth was the unmoving center of creation, and that once we got away from Earth, where there is change and decay, everything in the heavens was perfect and unchanging. Thus, Galileo's paradigm told him that the Moon had to be a perfect crystalline sphere and the Sun a perfect spherical ball of fire.

But that's not what he saw. On the Moon, he saw mountains, craters, and other "imperfections," while on the Sun, he saw a series of blemishes that we would call sunspots (or, more likely, sunspot groups). Most telling of all, when he looked at Jupiter, he saw four "stars" (his term) that were clearly orbiting the giant planet. Whereas his paradigm told him that everything in the heavens had to be circling Earth, his data were telling him that some objects seemed to be perfectly happy orbiting Jupiter. These bodies are now known as the Galilean moons, though Galileo, in an attempt to win courtly favor, called them the "Medicean stars." (The ploy apparently worked, because he was subsequently given a position in the court of the Medicis in Florence.)

Galileo published his findings in 1610 in a book called *Sidereus nuncius* (Starry messenger). It seems strange to the modern reader that the book caused a stir, since it is basically just a catalogue of what he saw through his telescope. But because he wrote in Italian and was therefore reaching the educated population outside the church, the book made him enemies—enemies who later would try him on suspicion of heresy for a subsequent book. We don't have room to go into that (somewhat convoluted) story here, although we should point out that Galileo's abrasive personality probably didn't gain him many friends, either.

In any case, once the new Copernican concept of heliocentrism was accepted, ideas about life in other locations in our solar system were quick to arrive. Some of the early thoughts about extraterrestrials seem pretty strange to us today. For example, in the eighteenth century, several serious astronomers proposed that there was life on the Sun—not on the fiery outer surface, of course, but in the (hypothetically cooler) interior. Some even tried to use their telescopes to peer through sunspots to see farms and villages below. And then there was the English country parson Thomas Dick, whose 1837 book *Celestial Scenery, or the Wonders of the Planetary System Displayed, Illustrating the Perfections of Deity and a Plurality of Worlds* confidently predicted that we would find 8,141,963,826,080 people living on the rings of Saturn. (From the tone of the book, we infer that the learned cleric expected them to be Englishmen.)

As the nineteenth century flowed into the twentieth, thoughts about extraterrestrials began to concentrate on nearby worlds—the Moon, Mars, and Venus. There was an almost universal belief that the Moon was populated—indeed, in 1901, H. G. Wells (1866–1946) published his novel *The First Men in the Moon*. In this book, earthly visitors wander about the lunar surface without space suits and encounter a technologically advanced race of

"Selenites." In 1895, the American astronomer Percival Lowell (1855–1916) published the first of three books about the planet Mars, titled *Mars* (1895), *Mars and Its Canals* (1906), and *Mars as the Abode of Life* (1908), in which he described his observations of the Red Planet. In these books, he developed a picture of an advanced civilization on Mars, a civilization that built a complex network of canals to bring water from the poles to the arid equatorial regions. He even reported the rate of poleward progression of vegetation with the seasons to three decimal places! Given that we now know that there are no plants or canals on the Martian surface, we have to wonder what Lowell actually saw when he looked into his telescope. The current consensus is that he was pushing his instruments past their performance limits and that he connected the random dots he saw into a pattern, much as people do when they take a Rorschach test.

Nevertheless, Lowell was a prominent figure in American science—the founder of Lowell Observatory in Flagstaff, Arizona—and his ideas influenced novelists and dreamers for decades. They probably also played a role in the promulgation of a strange kind of evolutionary theory of the solar system that was popular in the 1930s and 1940s. The basic idea was that Mars, Earth, and Venus represented progressive stages in development. Mars, arid and desolate, was the home of a dying civilization, whereas Earth held a civilization in full bloom. Venus—which was pictured as being like the Florida Everglades, only more so—was where civilization would flourish in the future. This beguiling notion, which is a perfect example of what the French call a *faux ideé claire* (clear but false idea), gave rise to all sorts of science fiction stories in which the swamps of Venus and the deserts of Mars played important roles. The notion that there were other civilizations out there was also involved in the development of a fictional character that one of the authors (JT) enjoyed as a child: Ming the Merciless, Emperor of Mongo, the villain in many Buck Rogers movies. (By the time these ideas were floating around, however, scientists had realized that the Moon had no atmosphere and no life.)

Today, we know that there is no dying civilization on Mars, and if life does exist there, it is (as we shall see in a moment) at best microbial. Venus, with a surface temperature hovering around 460°C (860°F), has no swamps or oceans and in any case is too hot to support the civilization of the future. Even though the notion of evolutionary progression had been dropped by the time serious exploration of the solar system began in the second half of the twentieth century, there was nevertheless a real hope that we would find life more or less like the life with which we are familiar somewhere in our neighborhood. When the Drake equation (see chapter 13) was first written down in 1961 to estimate the probability of making contact with extraterrestrials, for example, many scientists argued that life had developed elsewhere in our own solar system, and inserted their estimates of the number of times this had happened into the equation.

The Exploration of the Solar System

In a sense, you can read the last half of the twentieth century as a period of progressive restrictions on the places life could be found in our solar system. When *Apollo* 11 astronaut Neil Armstrong took his "giant leap for mankind" on the Moon in 1969, no one expected to find life on the lunar surface. Nevertheless, when the Viking spacecraft landed on Mars in 1976, there was a palpable sense of disappointment among both scientists and the general public that the onboard experiments showed no unambiguous evidence for the existence of life. Since then, a veritable flotilla of spacecraft from many countries has flown out to Mars, the most recent being NASA's *Curiosity* rover, which landed in 2012. These decades of exploration have established several facts about our nearest planetary neighbor:

• There were oceans on the planet early in its history.
• Evidence from the first flyby of the *Mariner* 4 spacecraft in 1964 showed systems on the surface that resemble terrestrial river networks, suggesting that there have been fairly recent upwellings of liquid water on the Martian surface.
• There is water ice in the upper Martian surface that sporadically appears on the outermost surface in liquid form.
• While there is no unambiguous evidence for the existence of Martian life, neither can it be ruled out categorically.

Given these facts, we suggest that the most likely scenario for Martian life is this: Life developed early on Mars, as it did on Earth. When the planet lost its ocean and atmosphere to space— after all, it is only one-tenth the mass of Earth—any life on the planet vanished. Today, the only evidence for this past life would be fossils on the Martian surface. This scenario, incidentally, explains NASA's obsession with mounting a "sample return" mission to bring back rocks that could hold evidence of that long-vanished experiment in biology.

But if explorations of nearby worlds seemed to restrict the possibilities for the existence of life, exploration of the worlds of the outer solar system quickly reversed that trend. With the launching of the Galileo spacecraft to Jupiter in 1989 (it arrived there in 1995), things changed. The reigning paradigm at the time said that the existence of life depended on the presence of surface oceans, as it does on Earth. We'll discuss this paradigm in more detail in the epilogue, when we talk about the so-called Goldilocks planet, but for the moment we'll just regard it as one more example of the operation of the curse of the single example, which we noted in the last chapter. We know of life developing only on our own world, where the nearby star keeps the surface temperature between the freezing and boiling points of water. Given this, we naturally expected to find no conditions friendly to life in the outer solar system, on worlds so far from the Sun. But things didn't turn out that way.

A word of explanation: When we talk about "worlds," planets come to mind first. In the outer solar system, this would mean the giant planets Jupiter, Saturn, Uranus, and Neptune. No one, however, expected to find life on a planet like Jupiter—the conditions there are just too extreme. On the other hand, it is easy to overlook the fact that each of these planets has many moons, each of which qualifies as a "world," with its own unique history and structure. (This, incidentally, will be an important point to keep in mind later, when we talk about planets circling other stars.) Indeed, the Galileo spacecraft was far too fragile to survive the crushing pressure it would have encountered had it tried to enter Jupiter's atmosphere (although it did drop a probe that returned data for a while). The surprises we found were not in the planet at all, but in its moons.

Although Galileo named his "stars" for the Medici family, the modern practice is to use mythology as a guide to naming. Thus, the moons of Jupiter were given the names of mythological characters who were—well, we guess the best term is "associated"— with Jupiter (whose Greek name was Zeus). According to one myth, Zeus became enamored of Europa, an aristocratic Phoenician woman, and changed himself into a white bull that mingled with her father's herds. When Europa climbed on the bull's back, he ran off and carried her to Crete, where she became a queen. Such is the legend that resulted in the name of the first of Jupiter's moons to send shock waves through the scientific community.

Europa was considered the second of the Galilean moons as far as distance from Jupiter is concerned—indeed, Jupiter II is one of its older names. (Recent discoveries of other bodies have moved Europa to the status of Jupiter's sixth moon.) It is quite small—less than 1 percent of the mass of Earth— and the smallest of the Galilean moons. It is also far enough from the Sun that its average surface temperature is –160°C (–270°F), well below the freezing point of water. Its small size means that it doesn't contain enough radioactive material to generate significant amounts of heat, so scientists expected the Galileo spacecraft to find a frozen, dead world.

It's not that there's no water on Europa. We knew that its surface consists of water ice, but at Europa's ambient temperatures this ice would be as hard as a rock. Appreciable fractions of the moon's surface are crisscrossed with cracks, but it also has large smooth areas.

The discovery that caused such a stir came from gravitational and magnetic measurements made by the Galileo spacecraft. The magnetic measurements indicated that there is a subsurface material capable of conducting electrical current. The small number of craters and the large smooth areas on the surface were also telling, because the other Jovian moons are pockmarked with the results of impacts. Since there is no reason why Europa shouldn't have been hit by meteorites as often as the other moons, the only conclusion we can draw is that there must be some mechanism that allows the moon to rebuild its surface over relatively short times (astronomically speaking).

All these phenomena can be understood in terms of Europa's structure. Under a covering of ice is an ocean of liquid water—an ocean whose volume exceeds that of all of Earth's oceans combined. The water is briny, most likely as a result of minerals leaching out of rocks, and is therefore a good conductor of electricity. This explains Galileo's magnetic measurements. Meteorite impacts would presumably crack the overlying ice, allowing liquid water to flow over the surface and cover the impact craters before freezing. This would account for the unexpected smoothness of the Europan surface. Finally, the large cracks are thought to be the result of tides in the liquid ocean, which exert enough pressure on the overlying ice to produce fractures.

The question of how thick the ice layer is remains a scientific problem, one exacerbated by the possibility that the thickness of the ice may vary from place to place. In some current models, for example, the thickness varies from 10 kilometers (6 miles) in certain places to as little as 10 to 100 meters (10 to 100 yards) in others.

We have already hinted at one problem with the idea of a subsurface ocean on Europa. Simply stated, how can you have liquid water on a body that is so cold and that has no conventional sources of heat? Discovering the solution to this conundrum opened new vistas in our thinking about possible abodes of life.

The answer has to do with Jupiter's gravitational effects on Europa. The moon is close in, completing an orbit in less than four days. The gravitational effects of the other moons guarantee that Europa is at a different distance from its home planet at different points in its orbit, and this, in turn, means that the solid structure of the moon is constantly flexed. Like a piece of metal that is bent rapidly back and forth, the solid structure of Europa is heated by this flexing. In fact, this so-called tidal heating is enough to keep Europa's subsurface ocean above the freezing point of water.

A dramatic confirmation of this picture occurred in 2013, when the Hubble Space Telescope detected a huge water plume being ejected from near Europa's south pole. The plume, some 200 kilometers (about 130 miles) high, was a sporadic event, something like a geyser in Yellowstone National Park, but it was certain evidence of the presence of liquid water on the moon. That same year, scientists going over old Galileo data found a region on Europa's surface where an asteroid impact had left some minerals usually associated with organic materials, which suggests that Europa, like Earth, may have had the molecular building blocks of life brought in by meteorites and comets.

All of this led to a rather surprising conclusion. Instead of being a cold, frozen world, Europa was suddenly promoted to being one of the primary locations for possible life in our solar system. Although no governmental agency has promised funding for a mission to Europa as of this writing, planetary scientists are already brainstorming about sending a space probe there and drilling through the ice to sample the ocean. Presumably the first step in such a mission would be to create a good map of the ice thicknesses to guide the choice of location for drilling.

Because of the possibility of life on Europa, the Galileo mission was terminated in 2003 by deliberately crashing the spacecraft into Jupiter to prevent any possible contamination of the moon by terrestrial microorganisms. Having said this, if life is indeed found in Europa's ocean, it will most likely be microbial. While such a discovery would release biologists from the curse of the single example, it would probably not generate a great deal of interest among the general public. One lesson we can take from this surprise discovery of a liquid ocean in our solar system's backyard, however, is that if advanced life forms do develop in a Europa-like environment, they are unlikely to develop a science of astronomy quickly. Because they couldn't see the stars in the normal course of affairs, they might have little interest in communicating with beings in other solar systems, at least until they had found a way to get through the ice and look out from their world's surface.

If Europa were the only moon with a subsurface ocean, we could write the phenomenon off to chance. But as early as 2005, the Cassini spacecraft in orbit around Saturn detected geysers spewing from the surface of Enceladus, one of that planet's

moons. Subsequent observations have shown that these eruptions are quite common—more than 100 have been seen—
and that the material coming out is liquid water mixed with
salt. It is thought, in fact, that eruptions from Enceladus supplied most of the material in one of Saturn's rings. In 2014,
the Cassini spacecraft actually flew through a geyser and determined that the material venting from the surface was briny
water, mixed with simple organic compounds.

The sixth-largest moon of Saturn, only about 480 kilometers (300 miles) across, Enceladus is another world that we would
expect to be dead and frozen. Like Europa, however, it is a ben-
eficiary of tidal heating, and it appears to have a liquid ocean in
its southern hemisphere about the size of Lake Superior. Current
estimates are that the ocean is about 10 kilometers (6 miles) deep
and covered by 30 to 40 kilometers (19 to 25 miles) of ice.

The way that the ocean of Enceladus was detected was
slightly different from the detection process for Europa. Rather
than measuring magnetic anomalies, the Cassini spacecraft
recorded a slight change in its velocity as it flew near the moon's
southern hemisphere. The basic point is that liquid water is
denser than ice—this is why the ice on a frozen lake stays at the
surface instead of sinking to the bottom. This difference in density manifested itself as a difference in the gravitational force the
moon exerted on the spacecraft, and this, in turn, caused Cassini's measured change in velocity.

Before leaving the outer solar system, we should talk briefly
about one of the other moons of Saturn, Titan. Titan is the largest moon in the solar system, bigger than the planet Mercury. It
is also the only moon in the solar system that has a significant

atmosphere. It is, however, very cold—surface temperatures hover around –180°C (–320°F)—and at this temperature familiar materials take on strange properties. Water ice, for example, is as hard as a rock, and methane (natural gas) is a liquid. In 2004, the Cassini spacecraft dropped a probe named Huygens (after the Dutch scientist Christiaan Huygens [1629–95], who first sighted Titan) that gave us our first close-up look at the moon's surface. The first reaction of planetary scientists to the information being beamed back was "Wow—this looks just like Earth." Indeed, there are long, Sahara-style dunes around the equator and lakes up near the poles. The problem is that these familiar geological formations are composed of unfamiliar materials. The lakes are liquid methane, which rains out of Titan's skies, and the dunes are made from hydrocarbons that also fall out of the skies—one researcher compared them to dunes made of coffee grounds.

Although the details of Titan's geology produced the sorts of surprises that scientists expect when they enter new worlds, there were no changes in paradigm associated with the exploration of Titan. As early as the 1980s, one of the authors (JT) reported the general expectation that when we got to Titan we would see the beginnings of the kind of organic chemistry that eventually produced life on Earth. And that, in essence, is what we found.

However, because of the extremely low temperatures on Titan, chemical reactions on its surface take place extremely slowly. You can get a sense of why this might be so by noting that when you put food into a refrigerator, your aim is to slow decay processes. Putting it into a freezer will slow these chemical processes even more. As a rough rule of thumb, the rate of chemical reaction is cut in half for every 10°C (18°F) drop in temperature. Thus, any organic chemical processes that take place on Titan would take place very slowly—so slowly that we might not recognize them as constituting life. This is a point to which we shall return later.

Water Everywhere

Think for a moment about the significance of what we have learned about the existence of water in the outer solar system. Traditional thought says that the presence of water is a necessary condition for the development of life. But our new discoveries seem to tell us that water is all over the place. Consider the following:

- Europa has a subsurface ocean with more water than is found in the oceans of Earth.
- Three of Jupiter's largest moons (Europa, Callisto, and Ganymede) have subsurface oceans—the fourth (Io) probably lost its water long ago because of its intense heat.
- At least one moon of Saturn (Enceladus) has a subsurface ocean, and, if our mathematical models are to be believed, so does Titan (although we have, as yet, no direct evidence for this claim).

Thus, even before we get out of the inner solar system, we find that water, far from being scarce in the universe, seems to be quite common. This is an unexpected result, at least as far as conventional thought is concerned.

Pluto and Beyond

Pluto was discovered in 1930 by a Kansan farm boy named Clyde Tombaugh (1906–97), who worked at the Lowell Observatory in Flagstaff, Arizona. His story is unusual. He was interested in astronomy and cobbled together a telescope from spare parts he picked up around the farm. He made some sketches of Jupiter and sent them off to the Lowell Observatory, hoping to get a useful critique. Instead, he got a job offer and, once he arrived in Flagstaff, was assigned the tedious job of looking for a planet beyond Neptune. Later in life, Tombaugh commented on this turn of events by saying, "Hell—it beat pitching hay."

At the time, people thought that there were some anomalies in Neptune's orbit caused by an as-yet-undiscovered planet farther out—a body they called Planet X. As it turned out, the orbital measurements of Neptune were incorrect, but the belief motivated the search for another planet, which Tombaugh discovered in due course.

Even though Pluto was considered to be the ninth (and presumably last) planet in the solar system for more than 75 years after its discovery, it always presented astronomers with problems. For one thing, it is small and rocky, and the theories we discuss below said it is in a place where gas giants should form. For another, its orbit is tilted with respect to the orbits of the other planets. In addition, Pluto actually spends part of its "year" closer to the Sun than Neptune does (it was last closest to the Sun in 1989).

Pluto's puzzles began to be resolved when astronomers started exploring the solar system beyond its orbit. It had been thought since 1951 that a disk of rocky debris, called the Kuiper belt, extended outward to a distance of about twice the orbit of Neptune. The general consensus, as we pointed out in the previous chapter, was that the belt was simply a ring of debris left over from the formation of the solar system. In 1992, astronomers at the Mauna Kea observatory in Hawaii documented the first objects in the Kuiper belt, but it wasn't until 2005 that the first real surprise showed up. In that year, astronomer Michael Brown and his colleagues announced the discovery of a body now named Eris—a body comparable in size to Pluto that orbits at the outer edges of the Kuiper belt. Since then, many more of these so-called Kuiper belt objects (KBOs) have been found, and some astronomers have suggested that Neptune's moon Triton is actually a captured KBO. Finally, in 2016, astronomers at NASA's Jet Propulsion Laboratory in California presented evidence based on the analysis of the orbits of some KBOs for the existence of a planet in the Kuiper belt that could be 10 times as massive as Earth.

With these discoveries, the true significance of Pluto began to become clear. It isn't the last of the planets, a lonely straggler at the edge of the solar system. It is instead the first of a rich trove of worlds orbiting far from the Sun. A few surveys of the Kuiper belt—think of them as pencil-thin probes through the disk—have led astronomers to believe that we will find thousands (perhaps even hundreds of thousands) of KBOs out there. Most will be merely smallish rocks, but many astronomers expect a dozen or more planet-sized objects to be found as well. If this prediction proves true, then the variety we found among the moons of the gas giants will have been only a prelude to the variety we will find when we explore this next part of our home system.

As we discuss later in the book, when the spacecraft *New Horizons* made the first flyby of Pluto in 2015, it found a world of unexpected complexity—a world that, like Europa, might even be an abode for life. As you read this, *New Horizons* has continued past Pluto to continue our exploration of the Kuiper belt.

How Did It Get to Be This Way? The Reigning Paradigm

Before we leave our familiar solar system, we need to take a moment to summarize both the paradigm that developed to explain how our own system formed and (remember the curse of the single example?) our conjecture about how other systems may have formed. Our ideas about the formation of the solar system go back to the eighteenth century, to the French physicist and mathematician Pierre-Simon Laplace (1749–1827), the author of what is called the nebular hypothesis. In this scheme, mutual gravitational attraction among materials in an interstellar dust cloud caused the cloud to contract, and during this contraction the rotation of the cloud increased, much as an ice skater's spin increases when she pulls in her arms. Eventually, this process led to a situation in which most of the cloud's mass was concentrated into a compact sphere in the center—a sphere that would eventually become the Sun—and a flattened disk of leftover material spun outward. (It was, incidentally, this theory that caused Napoleon to ask Laplace why he never mentioned God in his book on his theory. Laplace's famous reply, which may be apocryphal, was "Sire, I have no need of that hypothesis.")

This was basically as far as Laplace got, but today we know a lot more about the details of the formation process. We now understand that the contraction of the Sun continued until the temperature at its core got high enough to initiate nuclear fusion reactions and the energy streaming outward stabilized the newly born star. The effect of this turn of events on the proto-planetary disk was striking. Out to a place somewhere between the present orbits of Mars and Jupiter, the temperature became so high that materials such as methane and water remained in a gaseous state and were eventually blown out of the inner solar system by a massive flood of particles emitted by the Sun as it geared up. Beyond this so-called frost line—or, probably more correctly, ice line—these materials could solidify and be incorpo-rated into planets. Thus, the ice line marks the boundary between the inner and outer solar system.

So, the central feature of our paradigm became this: plane-tary systems should have small, rocky planets close to their stars, with larger gas giants located farther away. It is conventional to refer to the former as "terrestrial" planets and to the latter as "Jovian." This distinction seems to be based on such simple phys-ics that it's not surprising that it was firmly fixed in the minds of the first astronomers looking for exoplanets.

Throughout most of the twentieth century, scientists assumed that the planetary formation process was a relatively simple and sedate one, with the planets taking shape pretty much in their present locations by slow accretion. New com-puter simulations, however, have given us a different (and much more violent) picture of the events that led to our current roster of planets.

The inner terrestrial planets formed by the gradual aggregation of solid bits of matter (think sand grains) into larger objects called planetesimals, which could be anywhere from boulder- to mountain-sized. The planetesimals, through a continual process of collision, fragmentation, and aggregation, eventually formed larger bodies called protoplanets. The computer models tell us that as many as a couple of dozen Mars-sized objects caromed around the inner solar system in a titanic game of cosmic billiards, with some being ejected from the system and others combining into larger objects. Indeed, it was the collision of one of these objects with the nascent Earth that put the material that eventually formed the Moon into orbit. In addition, computer models tell us that in these early stages of formation, a complex interplay of gravitational forces caused Jupiter and Saturn to come closer to the Sun than they are now, sending many protoplanets into the Sun and ejecting others outward to the outer solar system, forming the Kuiper belt. Eventually, the system settled down, and the planets we see today swept up the remaining debris of the planetary disk and assumed their present appearance.

Computer models also tell us that the outer solar system underwent a similarly violent process of formation. After small, rocky cores had formed, as with the terrestrial planets, gravitational attraction allowed them to accumulate the hydrogen and helium that were still there, far from the Sun. Jupiter and Saturn apparently formed first, followed by Uranus and Neptune. As mentioned above, various complex gravitational interactions moved the outer planets around as they swept up debris, kicked material out of the asteroid belt, and eventually assumed their present orbits.

So that's our home system—a lot more complicated and diverse than scientists thought it was a few decades ago, and a lot more interesting. It seems that every time we explored a new region, we were surprised at what we found. This is a pattern that repeats itself as we turn our attention outward, beyond our own solar system.

3

A PLURALITY OF WORLDS

*There nowhere exists an obstacle to
the infinite number of worlds.*

Epicurus (341–270 BC)

There cannot be several worlds.

Aristotle (384–322 BC)

The question of whether there are solar systems besides
our own has preoccupied human beings for millennia,
even though the tools that have enabled us to find those
worlds are relatively modern inventions. For most of recorded
history, this question has been debated largely on philosophical
(or even theological) grounds, rather than in terms that we in
the twenty-first century would recognize as scientific. And as the
above quotes show, from the earliest times there were two competing schools of thought—one that held that other worlds like
Earth might exist and one that thought they could not.

It's important to remember that what we call science today requires hypotheses that can be tested—it is, in fact, the relentless testing of ideas against nature that distinguishes science from other forms of intellectual activity. Thus, for most of history, the search for what we now call exoplanets was not really science. Nevertheless, the quest for other planets has been addressed by some of the best minds the human race has ever produced, and some of the issues raised in these debates foreshadowed matters that scientists still wrangle with today.

Let's start with the Greeks. To understand what the Greeks were looking for when they approached the issue of what historian Steven Dick calls "the plurality of worlds," we have to understand how they saw the universe in which they lived. To them, it was an unquestionable fact that Earth, unmoved and immovable, sat at the center of creation. Around Earth, in a series of concentric crystal spheres, revolved the Sun, the Moon, the planets, and, in the outermost crystal sphere, the stars. Thus, when the Greeks spoke of other worlds, they were actually talking about other collections of concentric crystal spheres, each centered on its own central, immovable Earth. In their language, they were talking about other complete cosmoses (*kosmoi*). The Greeks' debate over whether such *kosmoi* existed—as we have already pointed out, there were two competing, deeply divided schools of thought on the question—resembled the early-twentieth-century dispute over the existence of other galaxies more than it did a search for other planetary systems.

Of the two schools, the one associated with the work of Aristotle was by far the most influential, since it dominated the thought not only of the ancients, but of medieval scholars as well.

In modern language, we would say that Aristotle came to his conclusions on the basis of his physics. To an Aristotelian, the world was composed of four elements: the familiar earth, fire, air, and water. Each of these had an innate nature that drove it to seek a particular spot in the universe. Something made of earth, for example, would seek out the center of the universe (which, to an Aristotelian, was the same as the center of the Earth), while something made of fire would seek out the periphery of the universe. Thus, left to themselves, objects made of earth would fall and objects made of fire would rise.

It's important to remember that an Aristotelian watching a rock fall would see not the action of the force of gravity, as we would, but an expression of the innate nature of the rock. The rock would be like a salmon swimming upstream, compelled by forces beyond its understanding to seek the spot where it was spawned.

Given this physics, it's easy to see why Aristotle rejected the notion of other worlds. Stated simply, if there were more than one Earth, how would a rock know which center to seek? How could it decide which way to fall? For Aristotle, the only way to avoid this dilemma was to assert that other Earths could not exist.

The competing viewpoint to Aristotle was that of the atomists—a minority opinion throughout most of history. Their view was that the universe consisted of an infinite number of atoms (a word whose Greek root, *atomos,* translates as "that which cannot be divided") separated by voids. The atoms were in constant flux, and the familiar cosmos resulted from the coming together of a group of atoms more or less by chance. In an argument that has similarities to those heard in modern debates about extraterrestrial intelligence (see chapter 13), the atomists argued that in an infinite universe, such comings together of atoms must occur repeatedly, and thus there had to be a plurality of worlds.

We stress again that the important thing about these ancient debates is that neither side made any statement that could actually be tested, so they were not really scientific arguments in the modern sense. Nevertheless, they set the tone for the debate as it resurfaced in medieval Europe.

Greek knowledge came to Europe circuitously, with works translated first into Arabic at places such as the House of Wisdom in Baghdad and then, after the Crusades, into Latin at places such as Toledo in Spain. Aristotle's main work on cosmology, *De caelo* (The heavens), was translated into Latin in about 1170 and was quickly incorporated into the curriculum at the new universities at Oxford and Paris. Thus, when Thomas Aquinas (1224–74) approached the problem of the plurality of worlds in his great quest to reconcile faith and reason, he was heavily influenced by Aristotelian physics. To Aristotle's arguments against the existence of other worlds, Aquinas added a Christian gloss, essentially arguing from the unity of God to a unity of God's creation.

This argument, however, quickly ran into theological problems. In essence, theologians argued that it limited the power of God and questioned His omnipotence. In 1277, the bishop of Paris, Étienne Tempier, apparently acting at the behest of Pope John XXI, produced a list of 219 beliefs that were to be considered heretical. This list, known as the Condemnations of Paris, included the idea that "the First Cause cannot make many worlds" (item 34). The "First Cause" is God, and the point of item 34 is that to claim that God cannot make other worlds is to limit His power, and hence to commit heresy.

The list seems to have grown out of a conflict between theologians and the (more liberal) arts faculty at the University of Paris, centering on the latter group's enthusiastic embrace of the new Aristotelian learning. In any case, it changed the plurality-of-worlds debate, because in order to avoid problems with church authorities, new ways to deal with Aristotelian physics had to be devised.

The man who first did this was William of Ockham (1280–1347), a philosopher who is best known for developing the notion of Occam's razor, which holds that the correct answer to any question is likely to be the simplest one. He countered the Aristotelian argument given above by stating that, while the innate nature of the four elements would not change from place to place, the way that this innate nature was expressed could. He gave the example of two fires, one in Paris and one in Oxford. The innate nature of the fires would cause their flames to seek out different points on the periphery of the universe, one above Paris, the other above Oxford. Exchange the positions of the two fires, however, and the Paris fire will seek a place above Oxford and vice versa. In the same way, something made of earth could seek out the center of another world, not our own Earth.

The thing that will strike the modern reader about this long history of argument about the plurality of worlds is that no one seemed to consider the question of whether any of these plural worlds contained living beings. In fact, it wasn't until the fifteenth century that the German theologian and churchman Nicholas of Cusa (1401–64) raised this issue. "Rather than think that so many stars and parts of the heavens are uninhabited," he argued, "we will suppose in many regions there are inhabitants."

This notion was part of the cosmology of Giordano Bruno (1548–1600), probably the best known of the early advocates of the multiple-world picture. Bruno pushed the ideas of Copernicus further than anyone else had done, arguing that other stars, like our Sun, had their own planetary systems, possibly with life on those planets. Bruno was an unconventional thinker, and he espoused many doctrines in opposition to accepted church teaching, including denial of the Trinity, the divinity of Christ, and the concept of transubstantiation. It was these theological views, rather than his cosmology, that led to his trial for heresy. In any case, he was burned at the stake as a heretic for holding such beliefs in 1600.

Once the idea of extraterrestrial life had been raised by Nicholas of Cusa, serious theological problems quickly followed. Philip Melanchthon (1497–1560), a Protestant theologian and collaborator of Martin Luther, argued that the idea of multiple worlds had to be rejected because of the theological issues it created. Christian doctrine (Catholic, Protestant, and Orthodox) is based on two important events: the Fall (characterized by Adam and Eve's expulsion from the Garden of Eden) and the Redemption (characterized by the crucifixion and resurrection of Jesus). If there are living beings on other planets, questions—debated today in the relatively new field called exotheology—must be asked. For example, did the Fall occur on every planet and for every race? If it didn't, was the Redemption needed for beings who had never experienced the Fall? If the Fall is universal, did Jesus have to go to every world to die and be resurrected, or were the events on Earth enough to cover everyone? If so, why is Earth so central? Are there other paths to redemption on other worlds? It's not hard to see how this sort of theological questioning could go on forever.

In fact, there is some historical precedent for dealing with the problems associated with extraterrestrial beings. When European explorers first encountered people in the Western Hemisphere, for example, Native Americans probably looked as strange to them as an extraterrestrial would look to us. In 1537, Pope Paul III issued an edict titled *Sublimus Dei* that said, in effect, that the people in the Americas had souls and should be converted to Christianity and baptized. More recently, Pope Francis brought this tradition of acceptance into modern times by stating that he would baptize a "Martian" if the opportunity arose.

With the Enlightenment and the beginnings of modern science, the plurality-of-worlds debate changed yet again. Important figures such as René Descartes (1596–1650) and Christiaan Huygens published cosmologies that explicitly showed planets circling other stars. And although their ideas, like those of the Greeks and the medieval philosophers, were not subject to observational verification, they were nonetheless based on the new physical principles that were then being discovered and sound comfortably familiar to the modern ear.

And so while philosophers and scientists continued to speculate on the existence of exoplanets and extraterrestrials—there was, as we have pointed out, a lively debate on the question of life on the Moon—it wasn't until the mid-twentieth century that technology advanced to the point that the whole issue could be taken outside the realm of philosophy and made part of observational astronomy.

Let's take a moment to discuss why finding exoplanets is so difficult. It wasn't until 1838 that the German astronomer Friedrich Bessel (1784–1848) was able to measure the distance to another star (for the record, it was the star 61 Cygni, 10.9 light-years away). For the first time, scientists had a sense of the vastness of the Milky Way, the incredible distances that separate stellar systems. (For reference, if we shrank the distance between Earth and the Moon down to a foot, the nearest star—Alpha Centauri—would be somewhere in the middle of Russia.) Thus, we are looking for relatively small objects very far away—always a difficult task.

Furthermore, while stars generate their own light, planets shine only by reflection, a fact that means that they are not only far away but very dim. One author compared the problem of finding them to the difficulty of detecting a birthday candle at the edge of a searchlight in Boston using a telescope located in Washington, D.C. Yet, in a situation like this, the dimness of the planet is not the main problem—the Hubble telescope has detected far fainter objects in the sky. The problem is that the planet is close to its star, which is an extremely bright light source. Separating out the light reflected from a planet from the glare of its star is especially difficult.

Faced with these twin obstacles, twentieth-century scientists began to search for exoplanets not by trying to observe the planets themselves, but by trying to observe the effects the planets have on their parent stars. To understand the techniques that were developed, it will help to imagine that you are an astronomer on a planet circling a distant star and that you are trying to detect the presence of planets in our own solar system.

We usually describe our solar system in a loose way by saying that the planets circle the Sun. This isn't a precise description, however—in fact, the Sun and a planet such as Jupiter circle a point between the two, called the center of mass. (We use Jupiter as an example because it is the largest planet and has the largest effect on the Sun.) You can think of the center of mass as the point where you could balance the two masses if they were connected by some impossible rod. Thus, while Jupiter completes its 10-year orbit around the center of mass, the Sun describes a much smaller circle in the same time frame. Since the center of mass of the Sun–Jupiter system is actually inside the Sun, the motion of the Sun is more like that of an out-of-balance washing machine than that of a planet, but that motion betrays the presence of the planet.

The first scientists trying to locate exoplanets looked for this motion directly. The basic strategy was to find a nearby star that moves through the sky quickly and then to look for wobbles in the star's path caused by a planet. In the late 1930s, astronomer Peter van de Kamp (1901–95) of Swarthmore College began observing a small object about five light-years from Earth known as Barnard's Star. Analyzing thousands of photographs taken over the next 40 years, he claimed to have detected the telltale wobble caused by a planet. Unfortunately, other astronomers have not been able to confirm his results, and it is now thought that the technical difficulties Van de Kamp faced in making these precise measurements were too large to be overcome with the technology available to him. Consequently, the consensus today is that he did not detect an exoplanet.

A second, more promising way to detect stellar wobble is to use the Doppler effect. This effect occurs when a wave such as light or sound is emitted by a moving source. If the source is moving toward an observer, he or she perceives the wavelength to be shorter by the amount the source moves between the emission of crests. In the case of sound, this means a higher pitch; in the case of light, this means a shift toward the blue end of the spectrum. Similarly, if the source is moving away from the observer, he or she perceives a longer wavelength—a lower pitch for sound and a shift toward the red for light. It is the Doppler effect that causes the familiar change in pitch in the sound of an automobile's horn when it passes you on the street.

Consider a star with one planet in orbit, and, for the sake of argument, suppose that the plane of the planet's orbit lies in the line of sight toward Earth. Then, as the star and its planet move around their center of mass, the star moves toward us for half of the revolution and away from us for the other half. This, in turn, means that the light we see is red shifted for a while and then blue shifted—in fact, we see a smooth oscillation of the wavelength between these two extremes. Because this technique measures the motion of a star toward and away from us, astronomers refer to it as a "radial velocity measurement."

The first discovery of exoplanets by this means was made by Aleksander Wolszczan and his colleagues at Penn State

University in 1992. The problem, as we pointed out in chapter 1, is that the two exoplanets they found are in a place where no one expected them to be. They are, in fact, orbiting a kind of star known as a pulsar.

A word of explanation about pulsars: when a large star burns through its nuclear fuel, it ends its life in a gigantic explosion called a supernova. The entire outer covering of the star is hurled into space while its core collapses into an incredibly dense, rapidly rotating object known as a neutron star. Typically around 16 kilometers (10 miles) across, the star emits a continuous beam of intense radio waves. These beams are like the beacon from a lighthouse, and if they happen to sweep across an observer, he or she sees a steady succession of radio pulses (hence the name *pulsar*).

The point is that the region of space around a pulsar is the last place you would expect to find exoplanets. Even if the star had a planetary system before it went supernova, it's hard to imagine how those planets could have survived the explosion. But in a paper published in the prestigious journal *Nature* in 1992, Wolszczan and his colleagues presented unassailable evidence that the telltale oscillation in frequency due to the Doppler effect is present in radio waves from pulsar PSR 1257+12 (PSR stands for "pulsating radio source," and the numbers indicate the position of the object in the sky). Not only that, but a detailed examination of the data showed not one but two planets in the system. No doubt about it—PSR 1257+12 has planets. Thus, Wolszczan (who, fittingly, had received his education at the Nicolaus Copernicus University at Torun in his native Poland) became the first astronomer to detect an exoplanet. Furthermore, the exoplanets are in a place that no exoplanet, in theory, has a right to be.

So the pulsar exoplanets around PSR 1257+12 became the first in a long string of surprises that the galaxy had in store for us. For the record, we still don't really know why they are where they are. Either they survived the supernova explosion (God knows how) or they formed out of the debris after it was over. Either way, they are strange beasts.

The first "normal" exoplanet was discovered in 1995. A team of Swiss astronomers working at an observatory in France, followed quickly by a team at the Lick Observatory in California, announced that there is a planet around the star 51 Pegasi (the name means that it is the 51st-brightest star in the constellation Pegasus, the flying horse), about 42 light-years from Earth. And again, the planet turned out to be a surprise.

This time, though, the surprise was of a different kind from that involved with pulsar planets. This new planet is in a normal stellar system, but it is big—perhaps half the size of Jupiter—and it is located close to its star, closer, in fact, than Mercury is to our own Sun. The problem is that there is no room for this kind of planet in the neat scenario we presented for our own solar system in the last chapter. Planets close to the star are supposed to be small and rocky, with large planets forming farther out. As the millennium turned, more and more of these so-called hot Jupiters turned up, and astronomers began to wonder whether any newly discovered exoplanets would be in systems like our own. Like the butterfly collector we talked about in chapter 1, we seemed to be finding cabbage butterflies where our old paradigm told us we should find monarchs. What to do?

As it turned out, a solution to this puzzle was not long in coming. With the launching of the Kepler spacecraft in 2009, it became clear that pulsar planets and hot Jupiters were just the beginning of our voyage into a strange new universe and the formation of a new paradigm.

4

WHAT IS A PLANET?

*I shall not today attempt further
to define [pornography].
But I know it when I see it.*

Justice Potter Stewart, Jacobellis v. Ohio, *1964*

S ometimes trying to create a hard and fast definition of
something is more trouble than it's worth. Mere words
are often a poor way to describe the complex set of cog-
nitive processes that go into recognizing an object, no matter
how common and ordinary that object might be. This is why Justice Stewart's comment on pornography is used so widely in the
intellectual community. In this chapter, we argue that it applies
equally well to the definition of a planet.

Having made this point, we acknowledge that defining a
"planet" has been made significantly more difficult by the discovery of the diversity and complexity of exoplanets—the existence
of hot Jupiters was the first example, and there will be many

more. Until recently, most of the debate on the nature of planets has been confined to the planets of our solar system and hence, in the end, suffers from the curse of the single example.

To the ancient Greeks, who first introduced the word *planet*, there was no ambiguity in its definition. In their world, there were the fixed stars whose relationship to one another never changed, and there were other lights in the sky that moved from place to place in a regular and predictable way. The latter were the planets (from the Greek *asteres planetai*, or "wandering stars"). Transitory phenomena such as comets were thought to be conflagrations in the upper atmosphere and therefore weren't part of Greek cosmology.

It's important to realize that to the Greeks, Earth was not a planet. Instead, it was the unmoved and unmoving center of creation, around which everything else moved. Technically speaking, however, the Sun and the Moon would have been classified as planets in the Greek cosmology since, like everything else, they moved around Earth.

After the Copernican revolution, the Sun moved to the center of the solar system in accepted cosmologies and the Moon became a satellite of Earth. In this scheme, there were six planets—the familiar Mercury, Venus, Earth, Mars, Jupiter, and Saturn. There was no debate about what was and what wasn't a planet, simply because all of these objects had been known from time immemorial. William Herschel (1738–1822) discovered Uranus in 1781, but this event, along with the subsequent discovery of Neptune in 1846, didn't really trigger any debate, since they seemed to engage in the same sort of behavior as the known planets—they were just farther away.

Actually, the naming of the planet Uranus tells us something interesting about how the astronomical community used to work. At the time of the discovery, Herschel was a professional musician and an amateur astronomer living in the town of Bath, England. (A second oboe in the band of the Hanoverian Guards in his native Germany, he had emigrated to England when he found military life unappealing.) A skilled telescope maker, he discovered the planet while making a systematic survey of the sky from his backyard. Following the example of Galileo and the moons of Jupiter, Herschel suggested the name Georgium Sidus ("George's Star"), after the king of England, for his new discovery. (The gambit seemed to work as well for him as it had for Galileo, since Herschel received a royal pension from George III.)

His suggestion, however, was largely ignored by his fellow astronomers, particularly those outside England. By the end of the eighteenth century, common usage had made the mythological name Uranus (the Greek god of the sky and grandfather of Zeus) pretty much universally accepted. A similar (though more muted) phenomenon followed the discovery of Neptune in 1846, when Urbain Le Verrier (1811–77), one of the astronomers who had predicted the planet's existence, pushed to have the planet named Le Verrier's Star. Unsurprisingly, outside France this name never caught on, either.

But in spite of these mild conflicts over naming, there was no question about whether what had been discovered was actually a planet. Unfortunately, this friendly state of affairs depended on the relatively primitive nature of the technology available to astronomers at the time. By 1801, however, telescopes had gotten good enough to detect Ceres, the largest body in what would later be identified as the asteroid belt. Eventually, many more such objects were detected. What to do? Does every rock circling the Sun count as a planet? If not, where do you draw the line? In the early nineteenth century, newly discovered asteroids were simply counted as new planets—indeed, even as late as 1867, one author claimed that there were no fewer than 90 planets in orbit around the Sun.

Over the last half of the nineteenth century, however, something interesting happened. There were no international bodies to take on the role of terminology police, but descriptions of the solar system began to change and evolve toward our current understanding. The asteroid belt was recognized as a system of bodies ranging from boulders up to Ceres, which is about 970 kilometers (600 miles) across. Without getting into a debate about what a "planet" is, astronomers apparently realized that most asteroids didn't belong in the category. Like Justice Stewart, they knew it when they saw it, and asteroids definitely weren't it.

The Classification of Pluto

To understand the current controversy about whether Pluto is a planet, we have to talk a bit about how the scientific community is organized today. At the national level, there are professional organizations representing different scientific disciplines. The authors, for example, are an astronomer (MS) and a physicist (JT) and, as such, are members in good standing of the American

Astronomical Society and the American Physical Society, respectively. Other professionals may belong to the American Chemical Society, the American Medical Association, and so on. One important organization that spans all the sciences is the American Association for the Advancement of Science (AAAS), headquartered in Washington, D.C. These organizations fulfill many important functions. They publish scholarly journals to report on new research, organize professional conferences, and generally represent the interests of their constituencies to the public and to the government.

In addition to these national scientific organizations, international ones began to be formed in the nineteenth century, often by mutual agreement among national groups. In 1875, for example, the International Bureau of Weights and Measures was set up by treaty. Headquartered near Paris, it defines and oversees the maintenance of important standards such as the second and the kilogram. It works in conjunction with national organizations such as the National Institute of Science and Technology in the United States to make sure that scientists and engineers worldwide have reliable standards for their measurements. Similarly, the International Union of Pure and Applied Chemistry (IUPAC), founded in 1919, has, among its duties, the task of standardizing chemical nomenclature. One example: when element 112 was produced, the discoverers wanted to name it after Nicolaus Copernicus (it is customary to name new elements after famous scientists—einsteinium, curium, and so on). Their proposed symbol, Cp, turned out to be already assigned to a group of chemical compounds, however, so the IUPAC changed the symbol to Cn.

When the International Astronomical Union (IAU) was founded in 1922, then, it joined a large and growing body of international scientific organizations. Its first task was typical of the sorts of things these organizations do. Astronomers usually identify stars by the constellation in which they are found—the star 51 Pegasi, as we explained in the last chapter, is the 51st-brightest star in the constellation Pegasus. As telescopes got better and better, and fainter and fainter stars entered the astronomer's purview, it became increasingly important to know where one constellation began and another ended. Other than vague artists' drawings, however, these boundaries were not well defined. The first job IAU astronomers had to tackle, therefore, was to survey the skies and define clear, permanent boundaries among constellations, a job they did very well.

Maintaining these kinds of international standards is a difficult and often thankless task, and we have nothing but respect for the men and women who carry it out. Theirs is important work, absolutely necessary for the maintenance of the scientific enterprise. None of it is likely, however, to make tomorrow's headlines or to serve as grist for late-night comedians. Every once in a while, though, one of these organizations makes a decision that is so colossally, mind-blowingly silly that you can only stand in awe when you see it. Examples: the decision by the International Commission on Zoological Nomenclature to rename *Brontosaurus* in 1999 and, yes, the 2006 decision by the IAU to "demote" Pluto.

So, let's get down to the basic question: is Pluto a planet or not?

The answer to this question depends on how you define the word *planet*. If you consult the *Oxford English Dictionary*, you find the word defined in its astronomical sense as follows: "any of the various rocky or gaseous bodies that revolve in elliptical orbits around the Sun and are visible by reflected light, especially each of the nine major planets." A definition like this wouldn't have been much help to the astronomers trying to deal with the asteroid belt in the nineteenth century, and it won't help us much in talking about Pluto. We need a lot more technical detail.

Consequently, the IAU's board, in anticipation of dealing with this problem at its August 2006 meeting in Prague, created a committee to come up with a workable definition of the word *planet*. The committee was headed by Harvard University historian of science Owen Gingerich. Gingerich is well known in the scientific community—he is one of the most respected astronomical scholars (the authors would say *the* most respected astronomical scholar) in the world. If anyone could deal with the complexities associated with Pluto, the IAU elders most likely argued, Gingerich was the man to do it. At meetings that Gingerich described as involving "vigorous discussions of both scientific and cultural/historical issues," the committee formulated the following definition: "A planet is a celestial body that (a) has sufficient mass for its self-gravity to overcome rigid body forces so that it assumes a hydrostatic equilibrium (i.e., round) shape, and (b) is in orbit around a star, and is neither a star nor the satellite of another planet."

Sensible. It's a planet if it's not just a rock and not big enough to be a star. What could be simpler? This definition certainly satisfies the "I know it when I see it" criterion.

Let's look at these requirements in a little more detail. If you look at a rock, you'll see that, in general, it has an angular shape. This is because the electrical forces among atoms in the rock are strong enough to overcome any other forces that might be acting on it, so the rock retains its angularity. Start making the rock bigger and bigger, however, and eventually a new force begins to enter the picture—the force of gravity. When your rock becomes hundreds of miles across, the gravitational force exerted by all that mass becomes big enough to overcome the electrical forces between atoms, and the material in the rock starts to respond to gravity. The atoms start to rearrange themselves in response to the force of gravity, assuming the round shape required by the Gingerich committee's first criterion. Thus, the asteroid Ceres (about 970 kilometers or 600 miles across) meets this criterion, but the random rock in the asteroid belt does not.

In the world of exoplanets, though, this criterion may need to be expanded a bit. For example, if a planet is rotating, a third force comes into play in addition to the electrical and gravitational forces discussed above: centrifugal force. This force takes the round shape and smears it out at the equator, so that the body appears to be slightly squished (the technical term is *oblate*). Even a relatively sedate planet such as Earth exhibits this effect, since the diameter from pole to pole is about 40 kilometers (25 miles) shorter than the equatorial diameter. Since this amounts to a deviation of only about 0.3 percent from sphericity, it's almost always reasonable to treat Earth as a sphere. For Saturn, however, this deviation grows to almost 10 percent of the planetary diameter, an effect not so easily neglected.

There are sure to be rapidly rotating exoplanets out there, so, in the future, what we mean by "hydrostatic equilibrium" may have to take account of centrifugal force. In 2014, in fact, one of the authors (MS) and two of his students (Prabel Saxena and Peter Panko) worked out the details of what the transit of a football-shaped planet would look like. In addition, we can imagine other, as yet unanticipated forces that might have to be taken into account someday—magnetism comes to mind, for example. Thus, as is always the case with exoplanets, we have to keep an open mind about what else we might discover and how it might affect our definition of a planet.

How about the other criterion—the one about not being a star? As the mass of an object gets bigger and bigger, the temperature at its center increases until, at some critical mass, the particles at the core are moving fast enough to initiate fusion reactions. When this happens, the object becomes a star. Thus, the criteria given above define a kind of middle ground between a rock in orbit and a double star system. The downside of this definition was that it would classify yet-to-be-discovered Kuiper belt objects (KBOs) as planets. When the definition was proposed, it would have included 12 objects (including Ceres) and opened the door to many more possibilities.

This potential expansion made some astronomers uncomfortable, for reasons the authors find hard to understand. During the nineteenth century, after all, chemists kept finding new chemical elements, but no one objected to identifying them all as elements. There was no move, for example, to demote some to "dwarf" elements. If nature has produced over 100 chemical elements, the community said, so be it. (The number actually stands at 118 and counting today.) So what's the problem with a couple of dozen planets in the solar system?

To understand what happened in Prague, though, you need to understand something else about scientific meetings. The conference in Prague was scheduled to last 10 days. This is a rather long time for a scientific meeting. Even the annual meetings of the AAAS, arguably the most comprehensive scientific gathering in the world, last only for what is, essentially, a long weekend. Throw in travel time and the 2006 Prague meeting of the IAU required an attendance of almost two weeks, an unusually large time commitment even for academic astronomers in the summer. Consequently, although many astronomers attended the conference, many came for only a few days, when their particular research interests were in play, and left before the conference was over. On the last day, in fact, only about 400 people were still in attendance, a fact that is important in understanding the resolution of the Pluto debate.

The problems actually had started when the General Assembly opened. They began with objections from the floor that the definition of a planet produced by the committee hadn't been announced in advance—a reasonable complaint, even though the organizers had acted in this way in an attempt to avoid a media circus. Things went downhill swiftly from there, however, with, according to an observer, one participant "literally screaming" because a representative from his own narrow specialty hadn't been included on the committee. One person present at the meeting described the proceedings in an email as "astronomers behaving badly."

In any case, by the last day of the meeting a new definition had been cobbled together and was presented for a vote. In essence, it added a third criterion to those listed above. This was that a planet must have "cleared the neighbourhood around its orbit."

This is a strange requirement, since the definition of *cleared* was never given. For example, in 2012, the good citizens of Chelyabinsk in Russia learned that Earth is still in the process of clearing its neighborhood when a house-sized meteorite landed near their city. Does this mean that Earth isn't a planet? Surely not, but the event illustrates the danger of trying to deal with a difficult definitional problem in haste. In this case, the question of how cleared an orbit has to be to satisfy the new criterion was not addressed, leaving the definition essentially useless.

In any case, according to the IAU, there are now two new subcategories of planets: dwarf planets, a category that includes Ceres, Pluto, and any new KBOs that may turn up, and plutoids, a category that includes any dwarf planets outside the orbit of Neptune (remember that Ceres is in the asteroid belt).

The reaction to the IAU action was immediate and intense. In the astronomical community, there was a widespread negative response. Over 300 astronomers signed a petition opposing the new definition—a number that may well exceed the number of affirmative votes actually cast in Prague. There was a persistent claim that the vote was actually an expression of anti-American attitudes by European intellectuals—Pluto, after all, was discovered by an American. By contrast, Neil deGrasse Tyson, one of the world's leading explainers of science and a champion of Pluto's reclassification, blames the negative American reaction on the existence of the Disney character named Pluto, although the authors see this explanation as being rather far-fetched.

Several factors, in truth, contributed to this negative reaction. For one thing, the word *planet* has many linguistic and cultural connotations outside astronomy. The *Oxford English Dictionary*, for example, lists no fewer than half a dozen definitions of the word. The idea that a group of narrow specialists would assume that they had the right to define a word with such a wide meaning smacks of a kind of arrogance. In any case, the upshot of the IAU vote is that many astronomers—possibly a majority— question the wisdom of the IAU definition.

We close this rather discouraging discussion of Pluto with one final point to consider: if Earth were moved to the orbit of Pluto, it wouldn't, according to the IAU, be a planet either.

Defining "Planet" in the World of Exoplanets

The IAU had the enormous bad luck to get involved in defining the word *planet* just as the Kepler spacecraft, which we'll describe in the next chapter, was about to reveal the true complexity and diversity of planetary bodies in our galaxy. This newly understood complexity actually suggests a simple solution to the "Pluto problem" that would not generate much opposition. We already divide the inner planets of our solar system into two categories: terrestrial (small and rocky) and Jovian (gas giants). Why not just add a third category—call it "Plutonic"—for Pluto and the KBOs? Given that the diversity of exoplanets will force us to define many new planetary categories anyway, why not just start the process close to home?

In any case, we don't feel that the IAU definition is a useful way to approach exoplanets. Thus, in what follows, we adopt an extension of the more sensible definition proposed by the Gingerich committee—call it the "expanded Gingerich criteria"—that includes the possibility of forces besides gravity being involved. Taking into account all forces acting on a body, if that body is bigger than a rock and smaller than a star and is not a moon, then, as far as we're concerned, it's a planet.

We know it when we see it.

5

THE KEPLER SPACECRAFT

Things are seldom what they seem.

W. S. Gilbert and Arthur Sullivan,

H.M.S. Pinafore

There's no doubt about it—the launching of the Kepler satellite in 2009 changed our view of exoplanets forever. Most of the incredible diversity we've talked about, and explore more fully below, has been discovered by this single spacecraft. Indeed, while the Kepler was operating at full power, new planetary candidates were being announced at a rate equivalent to several a day (although the Kepler team actually announced the candidates in large batches).

The Transit Method

The Kepler spacecraft detects planets by a method that is easy to describe, if not necessarily easy to carry out in practice. It's called the transit method, and it works because of a simple fact: when a planet passes between an observer and a star, the observer sees the light from the star dim slightly, then return to normal as the planet moves on. A repeated pattern of such dimmings is a fingerprint suggesting the presence of the planet. As simple as this technique sounds, however, it encounters problems when actually implemented.

To understand some of these difficulties, imagine, again, that you are an astronomer on a distant planet observing our own solar system. The orbits of the planets in our system lie in a plane known as the ecliptic. If you, the distant observer, are also in that plane, then your instruments will see the characteristic dimming and brightening as the planets pass in front of the Sun. In this case, the planets will move across the Sun's equator, producing the longest possible transit.

But what if you are not in the plane of the ecliptic? To see what happens in this case, imagine tilting the ecliptic plane slightly upward. In this case, you will still see transits, but the planets will appear to pass higher up on the Sun and the transits will be shorter. Keep tilting the ecliptic and you will reach a point where the planets no longer appear to cross the face of the star and the transits will disappear altogether. A similar situation will follow if you tilt the ecliptic downward.

What this means is that the transit method depends critically on the orientation of the ecliptic plane in any system we observe, and this, in turn, means that there is a limited range of angles for that plane that will allow planetary detection around a distant star. Planets in systems whose ecliptics are outside that range will simply be undetectable. You can see this most easily by imagining that you, our hypothetical astronomer, are observing our solar system from a spot 90 degrees above the ecliptic (i.e., you're looking down on the system). In this case, your instruments will record no transits, despite the presence of many planets.

But suppose we astronomers here on Earth are lucky and do see a transitory dimming in a distant star. Does this mean we have found a planet? Not necessarily. There are many processes that can cause slight dips in a star's output. For one thing, stars naturally exhibit small up-and-down swings in their output, which is one of the biggest sources of ambiguity for astronomers using the transit method. For reference, even a staid star such as the Sun exhibits a natural variability of about 10 parts per million (ppm), which amounts to an uncertainty in the fifth decimal place in measurements of its light output.

In addition, we know that the Sun goes through a regular 11-year sunspot cycle. At the peak of the cycle, there can be many sunspots or sunspot groups on the solar surface. And although sunspots are actually bright regions—they appear dark only because they are cooler than the surrounding material— they do produce a slight dimming because they replace areas of greater brightness on the solar surface. If you, as our hypothetical distant astronomer, had only a single dimming to analyze, you could easily mistake a sunspot event for a planet.

In a similar way, a double star system could produce something that looked like a transit. Again, imagine observing such a system, in which two stars revolve around each other, from the plane of that revolution. When the two stars are widely separated, you will see a brightness representing the full output of both stars. When one star passes in back of the other, however, the total brightness of the system drops, producing an illumination curve that mimics a planetary transit. This is particularly true if one star is only partially shadowed by the other.

Finally, as precise an instrument as the Kepler is, there is a limit to its resolution. What this means is that all the light coming from a small area of the sky is treated as if it comes from a single source. You can imagine that such a situation is capable of producing confusing results. For example, suppose that there was an eclipsing binary star system whose line of sight to an observer almost (but not quite) passed near another star that was being observed, and suppose further that the difference in lines of sight from the binary and the star being observed was so small that the telescope treated all the light as if it came from a single source. In this case, the dimming in the distant binary would mimic a planetary transit in the foreground star.

As this short list shows, there are many ways—more than we have room to explain, in fact—to produce signals that could be mistaken for planetary transits. For this reason, the simple observation of a transit does not establish the existence of a planet. Instead, an object exhibiting that characteristic light curve is designated as a Kepler object of interest (KOI) and subjected to exhaustive further testing, which we describe below.

The Kepler in Operation

Let's start with some basics. The Kepler spacecraft had a mass of about 1,000 kilograms (2,205 pounds) on launch—about the same as a smallish car—with an instrumental payload of 478 kilograms (1,054 pounds). It is not in orbit around Earth but instead trails after its home planet as it orbits the Sun. (For reference, the spacecraft orbits the Sun in 372 days, which means that it is continually falling farther behind Earth.) There are several reasons for this seemingly strange choice of orbit. For one thing, if the Kepler were in orbit around Earth, the planet would block out part of the sky. For another, mass concentrations in Earth produce slight changes in the gravitational field—perturbations that would make aiming the satellite more difficult. The central feature of the Kepler search strategy is that it continuously monitors the light from about 145,000 stars. Thus, when a dimming occurs, it is recorded by instruments on the satellite and reported back to Earth.

As strange as it may seem at first, the Kepler is not pointed toward the heart of the Milky Way, the part of the sky that has the greatest density of stars. The reason, alluded to above, has to do with the resolution of the telescope. The density of stars in the plane of the Milky Way is so high that it would simply be impossible to isolate a single star against all the background stars. Instead, the telescope is pointed upward and out of the galactic plane. (For experts, we note that it is pointed toward the constellation Cygnus.) It is also pointed upward out of the ecliptic plane, thereby avoiding contamination of its signal by asteroids, KBOs, and sunlight. The small volume of space examined by the Kepler means that all the statements we will make later about the existence of planets in the galaxy are extrapolations from a small (but probably representative) sample.

The main working parts of the spacecraft are a large mirror (1.4 meters, or 4.6 feet, across) and an extremely sensitive camera. The Kepler camera records incoming light in pretty much the same way that your camera does: by reading the output of charge-coupled devices, silicon chips that turn incoming light into a small electric current. The sensitivity of the camera determines the success or failure of a mission such as Kepler. Kepler scientists report a sensitivity of about 30 ppm—essentially an accuracy at the level of five decimal places. For reference, the natural variability of the Sun is about 10 ppm, and the change in light output due to the transit of an Earth-sized planet across a Sun-like star is about 80 ppm.

The satellite was launched in March 2009 after several delays due to budget problems at NASA. Three months later, it was returning scientific data. The actual results of these measurements are discussed in the following chapters, but to get a sense of where things stand now, we need to understand an important aspect of spacecraft operation. It is crucial that scientists be able to point an instrument at a specific spot in the sky and keep it pointed there while measurements are being made. This, in turn, means that there has to be a way to orient the spacecraft in the emptiness of space.

This task is generally accomplished through the use of reaction wheels, which are essentially gyroscopes whose axes of rotation stay fixed once they start spinning. Thus, a reaction wheel defines a direction in space that the spacecraft can use as a reference to maintain its orientation. The Kepler satellite originally had four reaction wheels to provide redundancy once the spacecraft was in orbit. In July 2012, one wheel failed. The spacecraft can operate on three wheels, but this failure took away the redundancy in the original design. Then, in May 2013, another wheel failed and the spacecraft lost its ability to orient itself. Its engineers shut it down as they tried to find a way to proceed with the mission using two reaction wheels.

In what can be described only as an engineering miracle, they found a way to return the satellite to active, if restricted, duty. The new technique depends on the fact that the Sun emits a constant stream of particles known as the solar wind. This wind exerts a slight pressure on any object in its path. In an operation that one engineer described as analogous to keeping a rowboat pointed upstream in a river, NASA engineers were able to use this pressure to stabilize the orientation of the satellite with respect to the solar wind. The remaining two reaction wheels can correct the inevitable drift from this position, so the satellite can have stable operation, but only in this restricted orientation. The new program was labeled K2 "Second Light," and, in 2015, Kepler scientists announced the discovery of two Earth-like planets by the reborn satellite.

From Object of Interest to Planet

Before we move on to discuss these two new exoplanets, we'll point out again that the detection of a dip in the luminosity of a star just gets us to the label "Kepler object of interest." Verifying that the dip is actually a planet requires a long process of validation. The first step is to raise the stellar system from a mere object of interest to a candidate. Doing that requires a complex series of operations involving the telescope's electronics and photometer—operations that we won't describe in detail but whose end result must convince scientists that the dip is real and not just a glitch in the system. More importantly, though, astronomers at this stage spend a lot of time ruling out the kind of contamination by eclipsing binaries discussed above. Only when these operations are successfully completed can the system under observation move from object-of-interest status to candidate status. At this point, the system is typically referred to ground- or space-based telescopes for further study.

We have already touched on some of the reasons that the Kepler, once it has identified a candidate, turns the validation process over to other instruments. The system is set up to avoid false positives—the mistaking of a signal generated by other processes, some of which we've discussed, for a signal generated by a planetary transit. In addition, planets far from their star make transits over long periods of time, and these times may well be longer than the operational lifetime of the Kepler. Jupiter, for example, makes a transit every 10 years, so an instrument like the Kepler observing our solar system would see only one transit (at best) during its lifetime.

Confirming that the signal actually is produced by a planet can take a long time, and there are usually about three times as many candidates as confirmed planets in the Kepler output. (For reference, the number of confirmed planets passed 1,000 in 2014 and passed 4,000 in 2016.) The confirmation process is carried out by space-based observatories such as the Hubble Space Telescope or (more often) by ground-based telescopes. Once the Kepler has identified a star that might have planets, the confirmation process doesn't necessarily require a high-level instrument to carry out the validation process. For example, a 32-inch teaching telescope at the authors' home institution, George Mason University, in the suburbs of Washington, D.C., has been used to validate no fewer than eight Kepler candidates simply by observing regular transits. Once you know which star to look at, it doesn't take a lot of technological firepower to make these kinds of observations.

Depending on the details of the system being analyzed, astronomers can use a number of different techniques in the validation process besides observing repeated transits. These include:

- Radial velocity measurements. As we saw in chapter 3, the gravitational pull of a planet can impart a small back-and-forth velocity to its star, and this change produces a Doppler shift in the star's light. This shift can be measured.
- Transit timing variations. In systems with multiple planets, a planet's orbit is affected by the gravitational pull of the other planets. This produces small but predictable variations in the observed transits.
- Reflected light variations. Like Venus in our own system, exoplanets exhibit phases as they circle their star. This causes small, continuous variations in the light output of the system that may be big enough to be detected.
- Stellar deformation. Sometimes the Kepler can detect changes in the shape of a star as its partner completes its orbit. This technique is most often used to rule out systems in which the companion is a small star known as a brown dwarf.
- Finally, we note that in 2016 a new method of confirming Kepler results was introduced. A group headed by scientists at Princeton University introduced a method of analyzing Kepler results based on a technique involving computerized processing of incoming data. In their initial paper, they

reported 1,284 new exoplanets and rejected 428 candidates as false positives. We expect this kind of technique to become important in the future, when massive amounts of new data from new satellites will need to be processed.

The Kepler Saga

Before we go on to explore the new worlds the Kepler has found, we would like to make a short digression to describe the incredibly complex story of how the satellite actually came into existence. Merely having a good idea isn't enough to convince a massive bureaucracy such as NASA to invest major resources in a new project, no matter how compelling. Nothing illustrates this fact better than the tale of how the idea of measuring planetary transits turned into an actual machine in space.

As early as 1971, astronomers had proposed the transit method for planet hunting, and by 1984, NASA scientists William Borucki and David Summers (no relation to MS) had established that, in order to find Earth-like planets around other stars, we would need to observe those stars from space. As often happens in this sort of situation, NASA provided some modest funding to support the development of the instruments and technology needed for a transit mission. At this point, the main emphasis was on searching for Earth-type planets—what we called Goldilocks planets in chapter 1.

The central technical problem at this stage was the development of extremely sensitive and stable light detectors, or photometers. The reason for this emphasis is easy to understand. The photometers have to be sensitive because, as we pointed out above, the amount of dimming produced by a planetary transit is very slight. If the instrument reading fluctuates by more than this small amount—a phenomenon usually referred to as "noise"—then the transit signal will simply be lost. Similarly, if the reading of the instrument drifts over time, we won't be able to compare different readings taken of the same star.

In 1992, NASA instituted the Discovery Program. Its goal was to send out a fleet of small spacecraft rather than a few large (and expensive) ones. NASA administrator Daniel Goldin referred to these missions as "faster, better, cheaper." The missions were supposed to cost less than $425 million and be completed in three to six months. The first proposal for a transit mission was submitted under the awkward title FRESIP (Frequency of Earth-Sized Inner Planets). The NASA review board rejected the proposal, mainly because it felt that the photometers available at the time were not up to the task of finding Earth-sized planets. Back to the drawing board.

In 1994, following work on the photometers, the group again submitted to NASA a proposal for a transit mission and was again rejected, this time because it was too expensive. In 1996, a modified proposal was submitted, incorporating reduced costs by changing the position of the satellite to its present orbit. This generated another rejection, this time because the team had not demonstrated that it could successfully monitor thousands of stars simultaneously. It was at this time that the mission name was changed to Kepler.

Throughout the mid- to late 1990s, a sample photometer was built and tested at Lick Observatory in California. Another proposal to NASA in 1998, and another rejection—this time because there were still doubts that the photometer was sensitive enough to detect Earth-type planets. In 1999, a system was designed and tested incorporating the sorts of noise that would be encountered in an actual mission. In 2000, Kepler was again proposed, and this time it made it onto a short list of missions to be developed. Finally, in 2001, it was accepted as a Discovery-class mission, and the actual building of the spacecraft began. Fifth time lucky!

The point of this excruciating story is not that Kepler had a particularly hard time getting NASA funding. Any mission this complex and expensive has to go through a similar process. What it shows is that you don't just have to be smart in this business—you have to be persistent.

The Kepler Universe

In chapter 3, we explained that the first discoveries of exoplanets were not what we expected. First, there are those pulsar planets—we still don't really understand where they came from, although current speculation suggests they formed from debris after their star exploded. Their discovery was followed by a bunch of hot Jupiters, planets bigger than Jupiter in orbits closer to their stars than Mercury is to our Sun. We also pointed out that the only detection method available at the time was the radial velocity measurement, which would pick up hot Jupiters before it picked up anything else. Thus, at the turn of the century, there was a strange sense of foreboding in the scientific community. Was our home system really such an oddball in the galaxy? Was there anything out there besides hot Jupiters and other planets that, although real, were nothing like Earth?

The launching of Kepler quieted those fears. The transit method isn't biased toward big planets, but it is capable of picking up dips in luminosity due to planets whose sizes range from smaller than Earth to bigger than Jupiter. It quickly became obvious that the hot Jupiters are actually a small fraction of the kinds of planets out there and that the "hot Jupiter problem" wasn't really a problem at all, but merely an artifact of the radial velocity detection method.

The detection of a transit allows us to determine a good deal about the exoplanet from some simple physics. The time between transits is the exoplanet's "year," and from this we can determine the radius of its orbit if we use standard astronomical techniques to estimate the mass of the star. The amount of dimming tells us how much light the exoplanet is blocking, which allows us to determine its radius and hence its volume. Measuring the radial velocity of the star as it and the planet revolve around the system's center of mass allows us to calculate the exoplanet's mass and therefore its density (which is just the mass divided by the volume), and this, in turn, tells us if we are dealing with a terrestrial-type planet (small, rocky, and dense) or a gas giant (large and gaseous).

We can often learn a great deal more about a newly discovered exoplanet through the use of spectroscopy. This technique depends on the fact that atoms of a specific chemical element emit and absorb a characteristic pattern of light known as a spectrum. It is sometimes possible to image the exoplanet directly by blocking out the light from its star. In this case, we can determine the composition of the exoplanet's atmosphere by directly observing the spectrum—the more light of a particular wavelength is emitted, the more of that particular element is in the atmosphere of the exoplanet. Even if we can't make this sort of measurement directly, we can observe the disappearance of specific light patterns when the exoplanet moves behind its star, and thus we can deduce the composition of the atmosphere.

Once we have verified the existence of an exoplanet, once we know where to point our telescopes, we can learn a great deal about that exoplanet through simple observations and calculations. The diverse list of planets given in chapter 1 was derived from this sort of work.

Visiting the Exoplanets

In the chapters that follow, we explore the new Kepler universe by imagining what it would be like to get on a spaceship and make a visit to each new world. We begin each visit by talking about what we can learn about the planet by observing it, and then try to explain those observations in simple terms. This process involves a certain amount of guesswork, as you might imagine, but no matter how strange our new planet may seem, the same basic laws of physics and chemistry that operate on Earth apply there. This means that our speculations can be based on a solid bedrock of scientific knowledge, even if applying those laws leads us to describe a world very different from our comfortable Earth. In addition, as we saw in chapter 2, we already have a lot of experience with new kinds of worlds from the exploration of our own solar system, an experience that will supply many useful analogies in our discussion. To remind ourselves about the speculative nature of all these scenarios, however, we include a section, appropriately titled "Caveats," at the end of each visit to talk about ways future measurements or theories might change the picture we're presenting.

Finally, we should say a word about how these exoplanets are named. The general scientific procedure is this: we begin with the name of the star, which is followed by a letter indicating the order in which the planet was discovered. Thus, for example, the first planet we visit is labeled 55 *Cancri* e, which means that it was the fifth planet discovered around the star 55 *Cancri*.

So fasten your seat belts and get ready to explore our marvelous new galaxy.

6

55 *CANCRI* E

DIAMOND WORLD

Twinkle, twinkle, little star
How I wonder what you are
Up above the world so high
Like a diamond in the sky.

Traditional children's song

The planet's whole atmosphere is glowing brilliant green and yellow, with a bit of red thrown in. The most intense light displays are
at the poles. Powerful radio bursts blast out from high above the surface. The planet itself is black and doesn't reflect light, despite the
intense heat that results from its close proximity to its star. An irreg-
ular network of gleaming white-hot lines interlace across the dark
surface, like some impossible Jackson Pollock painting. Here and
there, volcanoes spew something into the atmosphere that creates

a diamond-like brilliance, catching the light of the star in spectacular displays.

What is this place?

The 55th-brightest star in the constellation of Cancer (the Crab), 55 *Cancri* is a pretty ordinary star: a little smaller than the Sun, a little cooler, a little over 40 light-years from Earth. At first it doesn't seem to have much to command our attention, though there are a few intriguing characteristics we might mention. The star is, for example, much older than the Sun—between 7.9 and 8.4 billion years old, in contrast to the Sun's comparatively youthful 4.5 billion years. It also has a somewhat higher concentration of heavy elements. It isn't, however, the properties of the star that brings us to this section of the sky, but the fact that 55 *Cancri* is surrounded by a complex planetary system that includes the body 55 *Cancri* e, a planet first detected in 2004. This is the place we are calling Diamond World, one of the most extraordinary planets discovered by the radial velocity method.

Stars and planets are made from materials in the same interstellar cloud, so we would expect the high concentration of heavy elements in the star to be reflected in the composition of its planets. One way to think about the role of stars in the universe is to picture them as huge factories that use primordial hydrogen as a fuel for nuclear reactions, producing "metals" that are returned to the cosmos when the star dies. (In the jargon of astronomers, any element heavier than hydrogen is referred to as a "metal"— even gases such as oxygen and elements such as carbon.) These metals are then taken up in new stars and planetary systems when they form. Our own Sun, for example, can be thought of as a third-generation system, having been formed from material created in three successive stellar cycles. 55 *Cancri* is most likely a fourth-generation star—a fact that explains its higher content of metals.

Hypothetically, as we approached 55 *Cancri* in our imaginary spaceship, our first encounter with its planetary system would have occurred when we were about 100 times as far from the star as Earth is from the Sun. (Astronomers refer to the Sun-to-Earth distance as the astronomical unit, or AU, and we adopt this nomenclature in what follows.) We'd run into a disk of icy and rocky materials ranging from asteroid-sized to the size of Earth's Moon. This outer disk is a direct analog to our own Kuiper belt (see chapter 2). Moving inward, we would encounter our first large planet, labeled 55 *Cancri* d, located 5.8 AU from its star. This planet is about 3.8 times as massive as Jupiter.

The rest of the planets in this system are located less than 1 AU from their star, but before we get to them, we'd pass through another familiar feature—an asteroid belt. Made up of the same kind of rocky material we see in our own asteroid belt, the belt around 55 *Cancri* would have been thinned out by gravitational interactions over the long history of the system, and those same interactions would have produced the kind of gaps in the belt that we see in the rings of Saturn. Most likely it would have hundreds of small gaps, making it look something like the grooves on an old-fashioned vinyl record.

Once we got to 0.78 AU—closer to the star than Earth is to the Sun—we'd encounter the objects that brought us to this system: a compact set of four planets ranging in size from about eight times the mass of Earth to almost the mass of Jupiter. It is the innermost planet of this group, the world labeled 55 *Cancri* e, that will be the focus of our attention for the rest of our visit in this chapter.

The first thing we notice about the planet is that it has an unusual orbit—it's tilted almost 80 degrees with respect to the equatorial plane of the star. In our own solar system, Pluto has the most tilted orbit, and that tilt is only 17 degrees. More importantly, though, 55 *Cancri* e moves very fast. It completes an orbit in only 17 hours, which means that its "year" is less than an Earth day. (Most astronomers assume that the planet is "tidally locked," which means that it always presents the same face to the star, much as the Moon always presents the same face to Earth. This means that the lengths of the planet's "day" and "year" are the same.)

The basic laws that govern planets are called Kepler's laws of planetary motion; they were discovered by Johannes Kepler (1571–1630), after whom the Kepler satellite itself is named. If we apply these laws to 55 *Cancri* e, we find that its orbit lies within the chromosphere (outer atmosphere) of its star. In a real sense, the planet is actually *inside* the star!

We can start our analysis of the planet by trying to figure out what 55 *Cancri* e is made of. Its radius is about twice that of Earth, and it has a mass almost nine times that of our own planet. This means that it has a density similar to rocks and metals on Earth—about six times the density of water. The size and density of 55 *Cancri* e pose some problems. For one thing, a planet of this size in our solar system would lie in the outer reaches, like Uranus and Neptune. But these so-called ice giants have a density only about twice that of water. Furthermore, if a planet the size of 55 *Cancri* e had the same chemical composition as Earth, its own gravitational force would compress its rocks to twice the density we actually see. We have to do some careful calculations, but when we do, we find that we can understand the size and density of this planet if half of its material is carbon, with the rest being silicon, oxygen, and a smattering of other elements. This is our carbon world.

This actually explains why most of the planetary surface is dark. Because it is so close to its star, 55 *Cancri* e receives almost 3,000 times as much stellar radiation as Earth receives from the Sun. This means that the planet is extremely hot, with a surface temperature near 2000°C (3200°F). Nevertheless, the planet seems to be dark, with a reflectivity something like that of charcoal (which is, after all, another form of carbon).

We can begin to understand the bright lines and volcanoes by noting that the electromagnetic radiation we receive from the planet's surface isn't reflected starlight, as it would be if we were observing a planet in our own solar system, but energy forcing its way up from the interior of 55 *Cancri* e itself. The interior of 55 *Cancri* e must be very hot, and most likely contains liquid "metals." We'll see where this fact leads us in explaining the planetary surface.

The radio waves we detect are something we usually associate with particles running along magnetic field lines. In our own solar system, planets such as Earth and Jupiter generate magnetic fields through a combination of rotation and the presence of materials that conduct electricity in their interiors. Earth's magnetic field, for example, is generated by what is called a dynamo associated with the rotation of the liquid part of its iron and nickel core, while Jupiter's magnetic field is associated with the rotation of metallic hydrogen. Particles, mostly from the solar wind, are accelerated along the lines of the planetary magnetic fields, producing the northern and southern lights on Earth when the particles collide with molecules in the atmosphere. The same effect produces radio emissions in Jupiter's much more extensive magnetic field.

It is reasonable to suppose that 55 *Cancri* e should behave like Jupiter, only more so. Like all Jovian planets, it would have a rocky, metallic core several times the size of Earth. In fact, given its rapid rotation and the fact that it probably has conductive liquid material such as iron in its interior, we would expect to see a magnetic field several thousand times stronger than that of Earth. In addition, given the proximity of 55 *Cancri* e to its star, and assuming that the star emits something like a solar wind (as most stars do), we should also expect to see some pretty spectacular fireworks associated with the outflow of stellar particles.

The reason for this is that the system we've described involves two strong but competing forces: one (magnetism) tries to protect the planet's atmosphere, while the other (the stellar wind) tries to destroy it. We can start to understand this interplay by noting that geological processes in the planet—processes we describe below—are constantly releasing gases, perhaps carbon dioxide or hydrogen cyanide, into the atmosphere. The intense heat quickly breaks up these molecules into their atomic components. If there were no magnetic field on the planet, the strong stellar wind would blow this nascent atmosphere off as fast as it formed, leaving an airless ball behind.

The presence of a magnetic field changes all that, however. On 55 *Cancri* e, as on Earth, the magnetic field deflects the incoming particles in much the same way that a large rock deflects water in a swiftly flowing stream. The planet's magnetic field, which we would normally expect to be a symmetrical two-lobed structure, is distorted and stretched in the downstream direction. On the upstream side, pressure from the stellar wind compresses the field. And that is our competition—the stellar wind trying to push the magnetic field down to the surface, and the magnetic field pushing back to keep the particles away.

It could happen that these two forces come to equilibrium and produce a stable atmosphere. We suspect, however, that they will not for this system. We know that the solar wind from our own Sun is erratic, producing bursts we usually call "solar storms" or "space weather." These events often have serious effects on terrestrial communication systems and power grids and are a major source of danger to unshielded astronauts in space.

If the star 55 *Cancri* behaves in a similar way, we can imagine scenarios in which intense stellar bursts push the magnetic field down and blow away the atmosphere. Then, when the storm subsides, the magnetic field would reassert itself and the atmosphere would be rebuilt, growing in concentration until the next stellar outburst. The atmosphere in this case would be in a constant state of change.

Finally, the fact that the planet's magnetic field is distorted and stretched out downstream by the stellar wind leads to a phenomenon known as reconnection, whose main effect is to accelerate particles back toward the planet. Screaming in along the field lines, they would produce northern and southern lights so intense that you could read this book in the middle of the night if you were on the surface.

To understand the appearance of the sky on this strange world, however, we have to think a little more about the atmosphere with which these incoming particles are interacting. On Earth, the air we breathe is made up mostly of molecular nitrogen and oxygen—two atoms of nitrogen or oxygen bound together. On 55 *Cancri* e, however, the temperature is too high for molecules to stay together. In addition, the high temperature makes it easy for light atoms such as hydrogen and helium to escape into space. We expect, then, that whatever atmosphere the planet has is composed of atoms of carbon and oxygen being released from the surface as carbon dioxide molecules and then broken up as described above—mostly a carbon atmosphere for a carbon world. Given the known properties of carbon, we can understand the green and yellow colors we observe, since this is the kind of light that carbon atoms emit, while oxygen atoms supply some red and green as well. Furthermore, given the fact that the stellar wind particles are being funneled toward the poles by the magnetic field, we understand why those regions are brighter than the rest of the planet.

What about that network of bright lines on the surface? Again, an analogy with our home planet can help. We know that the surface of Earth is constantly changing—that no geological feature lasts forever. The main reason for this situation is that the interior of the planet is very hot, partly because of the decay of radioactive nuclei and partly because Earth is still in the process of cooling down after its fiery beginning. As a result, the interior of the planet "boils," moving around chunks of the surface known as plates. The continents ride on the plates, and so are in constant motion. This process, known as plate tectonics, should be seen on any planet where there is a significant heat flow from the interior.

On Earth, the boundaries of the plates produce well-defined geological features. The Mid-Atlantic Ridge, for example, is an undersea mountain chain extending from Iceland to Antarctica. It marks the place where hot magma from the interior comes to the surface and then cools to create new crust. Similarly, the Andes Mountains on the western coast of South America mark a place where one plate is pushed beneath another and crust is destroyed. These sorts of places are known as subduction zones and are often associated with volcanic activity. The history of Earth can be thought of as a constant interplay between the creation and destruction of surface material, with continents moving around in response to forces deep beneath the surface.

Because 55 *Cancri* e has a mass almost nine times that of Earth, the cooling-off process will take much longer there than it has for Earth. Consequently, we should see that same interior "boiling" process that we see on Earth. The deepest interior should, in fact, contain a material unknown on Earth—liquid carbon, or, if you prefer, liquid diamonds. (Remember that it is the rotation of this molten material that gives rise to the magnetic field described above.) The bright lines we see on the surface are, in fact, places like the Mid-Atlantic Ridge, where molten material from the interior is coming to the surface, glowing brightly until it cools off.

But the most spectacular geological features on 55 *Cancri* e could be the subduction zones and their associated volcanoes. As molten carbon from the interior is brought to the surface, it may crystallize into diamonds before being spewed into the atmosphere, creating a light show unlike anything we've ever seen on Earth. Coupled with the intense auroras, the diamonds would make the surface of this world an artist's dream—provided, of course, that we could find a way for the artist to survive there.

This carbon world is a good case study of a type of planet we expect to find in some abundance in the galaxy—large planets made primarily from heavy elements. We can, in fact, imagine many variations on the 55 *Cancri* e theme. We will probably find iron worlds, nickel worlds, sulfur worlds, and so on, since we know that all of these elements are made in stars. (In fact, the planet Mercury in our own solar system could reasonably be labeled a nickel-iron world.) Each of these heavy metal worlds would have its own peculiarities, its own unique characteristics. As always, the variety we find when we look at exoplanets challenges our imagination.

Caveats

First, we should point out that measurements of the density of 55 *Cancri* e are still in the process of being refined, so our estimates of the carbon content on this world may change. It's hard to see, however, how such refinements could alter the conclusion that the planet has a high percentage of that element.

The existence of a magnetic field and the behavior of solar wind particles are based on pretty solid reasoning, so unless

55 *Cancri* is one of those rare stars without a solar wind, that part of our description should stand.

The discussion of plate tectonics is a little more speculative. The laws of physics and our experience with our own solar system tell us that there must be an intense heat flow out of the planetary interior, but how this heat flow affects the surface is a somewhat trickier question. Theoretical studies of planets in our own solar system indicate that the presence and motion of plates depend critically on details such as the amount of water in the planet's interior. A planet without much water might find its surface motion locked up, like a frozen engine. We think, for example, that it is the low concentration of water on Venus that prevents it from exhibiting plate tectonics, even though it is roughly the same size as Earth. We argue, however, that it is reasonable to expect to find the kind of plate boundaries and diamond volcanoes we've described on 55 *Cancri* e. And if we don't find those features on 55 *Cancri* e, we'll undoubtedly find them somewhere else in the galaxy.

7

HAVEN

ROGUE PLANET

One is one and all alone
And evermore shall be so.

"Green Grow the Rushes, O,"
traditional English counting song

It's dark. Very dark. The only lights are pinpricks of illumination from stars located light-years away. But it's not cold. In fact, it's like a house in which the lights have been turned off but the furnace is still running. There's no sun in the sky, but you can hear waves lapping on a distant shoreline. Any life in this place must see in the infrared, because there is almost no visible light.

We must be on a planet, or at least a moon, but where is this planet's sun? Doesn't every planet have to be in orbit around a star?

Well, maybe not. Think back to our description of the formation of the solar system in chapter 2. We can think of the formation as proceeding in three stages, something like shifting gears in a car. First the planets form, then the remaining dust and debris get absorbed or thrown out of the system, and finally the planets begin a complex dance driven by the force of gravity—what we called a game of cosmic billiards. In this last stage, planetary orbits shift around and planets can be ejected from the solar system. According to some theoretical calculations, as many as a dozen Mars-sized objects may have been thrown out of our own nascent solar system during this phase.

However, these ejected bodies don't just disappear. They continue wandering around the galaxy, most likely in orbits around the galactic center. In fact, it is likely that there are many more of these "rogue planets" out there than there are planets circling stars.

Faced with this fact, our first impulse might be to think that rogue planets must be pretty uninteresting—frozen worlds with no life, no heat, no energy. We argue, however, that this impulse results from what we called surface chauvinism in chapter 1. We know that even on Earth, life is not confined to the surface. There are complex ecosystems around deep-sea vents, for example, and bacteria living in rock a mile beneath the surface. Both would do quite well if Earth itself became a rogue planet, although in that case the oceans would freeze over. In fact, as we said above, a planet without a star is a little like a house with the lights turned off but the furnace running—the inside temperature of the house (or of a relatively large planet) really isn't affected much by the star's absence.

In the previous chapters, we have come across four different ways that energy can be supplied to a planet:

1. Direct radiation from a star—this is what powers life on Earth's surface.
2. Cooling off from the initial period of formation, a process that can take many billions of years (it's still supplying energy to Earth, for example).
3. Radioactive decay of elements in the planetary interior— the decay of uranium is an important source of Earth's internal energy.
4. Tidal heating—this is what produces the subsurface oceans on the moons of Jupiter and Saturn, as we saw in chapter 2.

Looking at this list, we see that only the first item depends on the presence of a star. A planet wandering off on its own still retains its original internal heat, and radioactive nuclei still decay inside it. If the planet is a gas giant, it may well be accompanied by its moons, and the force of gravity still produces tidal flexing to sustain subsurface oceans just as it does on some of the Galilean moons of Jupiter. Except for the surface, then, the presence of a star is pretty irrelevant for any planet or moon, and we are as likely to find life on a rogue planet as we are to find it in the oceans of Europa or, for that matter, around deep-ocean vents on Earth. And this means that rogue planets are as interesting to study as their star-circling cousins.

Given this fact, we are immediately confronted with the question of how to detect rogue planets. Neither the radial velocity method nor the repeating transit method—the techniques we've used to find exoplanets up to this point—will work, because they both depend on the presence of a star. Alternatively, since (as we'll see) rogue planets generate a lot of heat, we could look for them with infrared detectors. Unfortunately, our current infrared detection sensitivity is too low to allow this to be a feasible means of finding these worlds at this time.

Instead, finding rogue planets will require a totally new way of trying to find planets, since up to now we have discussed only techniques requiring the presence of a star. In fact, astronomers have known of such a technique for some time. It's called gravitational lensing, and it comes from—of all places—the general theory of relativity.

The theory predicts that when a beam of light or other kind of electromagnetic radiation passes near a massive object such as the Sun, its path is deflected. Looking back along this deflected path, it appears to an observer that the light beam is coming from a different place than its actual source. It was, in fact, the detection of this deflection during a solar eclipse in 1919 that first propelled Albert Einstein to international prominence.

Now imagine a rogue planet moving into a position directly between you and a distant star. Imagine further a cone of light emitted by the star in such a way that the rogue planet is at the center of the cone's base. Each ray in the cone's surface is deflected by the mass of the rogue planet, and those rays come together at the point where you are making your observation. In effect, the rogue planet acts as a kind of lens, focusing the light from the distant star. Looking back along these focused rays, however, you see not a single point of light, but a ring of light corresponding to the extension of that focused cone. This so-called Einstein ring indicates the presence of a massive object between you and the star, although if the alignment isn't exactly as described, you'll see arcs of light rather than a complete ring.

Gravitational lensing is widely used in astronomy to detect massive objects of galactic size, especially concentrations of dark matter, but it can be used to detect rogue planets as well. The challenge, of course, is that we don't know when and where a ring will show up, so there is no monitoring technique that can maximize the chances of detection. The best we can do is depend on chance alignments to give us a ring now and then. For this reason, we have seen evidence for only a handful of rogue planets to date. From these few detections, however, astronomers have estimated that the number of rogue planets in the galaxy is larger than the number of planets circling stars. Because of the uncertainties in the data, these estimates are extremely imprecise, but they put the number of rogue planets between two and 100,000 times greater than the number of their more sedentary brethren.

Given the paucity of data, we are going to take a different approach to discussing rogue planets than we did for Diamond World. Instead of imagining a visit to a known world, we will imagine what the life history of a rogue planet might be. We'll name our fictional planet Haven for reasons that will become obvious below.

Metaphorically speaking, our own Sun is located in the far suburbs of the Milky Way. On a clear night, we can see about 2,000 stars in the sky. Move in toward the galactic center (still suburban but closer to "downtown") and you can easily get into a position where the skies would blaze with 100 times as many stars. We'll put Haven's birthplace in a region like that.

In chapter 6, we saw that all heavy elements ("metals") are made in stars and returned to interstellar space when stars die. Because Haven's star would have formed closer to the center of the galaxy, we expect that there would have been many more supernovae in its neighborhood than in our own, and hence that objects in Haven's home system would contain more heavy elements than are found in our own system. These heavy elements, of course, include radioactive materials capable of generating heat.

We expect that the basic process of planetary formation for Haven and its neighbors would have gone on pretty much as it did in our own system, with perhaps some variations due to the greater number of supernovae in the region. Inside the ice line we discussed in chapter 2, we find small, rocky "terrestrial" planets, with gas giants forming farther out. Like Jupiter, these gas giants have a rocky core, perhaps several times the size of Earth, covered by massive mantles of hydrogen and helium.

Oddly enough, this structure would introduce a new supply of heat into Haven's interior. At the kind of pressures we find inside the planet, liquid hydrogen and helium don't mix, so the heavier helium would form into drops and fall toward the planetary center. This helium "rain" would have the effect of releasing gravitational energy, adding yet another source of heat to the list we gave above. We know that this process occurs because we can detect its effects in Saturn.

The formation of our planet is in some ways a miniature version of the formation of its solar system. As the central body of Haven takes shape, it would acquire a disk of materials that form planetesimals and, eventually, moons. The location and size of these moons would be a matter of chance, but just for the sake of argument, let's suppose that Haven acquires four moons like the Galilean moons of Jupiter and one large moon—maybe an Earth-sized moon—farther out. We also expect that collisions between the planetesimals would supply enough rubble to populate a series of rings around the planet—rings similar to those we see around Saturn. Gravitational forces, over time, turn these rings into the "vinyl record" configuration we saw in the last chapter.

The fact that all these moons are forming beyond the ice line means that many of the planetesimals in orbit would contain water ice. Thus, as each moon forms and its temperature increases because of heat generated by infalling objects, it reaches a point where surface oceans form. Add in the extra energy provided by the decay of radioactive nuclei—nuclei that, as we saw above, could be more abundant than in our own solar system— and those oceans might even boil. Eventually, however, enough energy would be radiated into space to drop the surface temperatures of the moons below the freezing point, and the oceans would be covered with a layer of ice. From this point onward, the subsurface oceans would be kept liquid by a combination of the residual heat of formation, radioactivity, and tidal heating.

It's essential to realize that Haven's star has little to do with the formation of the moons. The most important force operating is the gravitational attraction of the planet itself. As the game of cosmic billiards progresses in Haven's system, we can imagine many scenarios that could result in the ejection of planets. For example, Haven might get locked into a gravitational tug-of-war with one of the other gas giants, a tug-of-war resulting in the ejection of Haven and one or more terrestrial planets and the rearrangement of the orbits of the remaining planets. Nothing in this process, however, would separate Haven from its moons—the mutual gravitation involved is much too strong. So when Haven heads out into the galaxy, leaving its star behind, it takes its moons along with it. As discussed in chapter 2, our current theories suggest that planets were ejected early in the history of our own solar system.

What happens to Haven after ejection depends on the details of the ejection process. The planet could go into a highly eccentric, comet-like orbit around the galactic center, or it could, like the Sun, settle into a sedate circular rotation, making a circuit every 200 million years or so. Whatever happens, however, Haven is extremely unlikely to suffer collisions with another star or rogue planet—space is just too empty for that to happen. It will, in fact, achieve a kind of immortality, circling around in interstellar space for billions of years as it slowly cools off.

While this is happening, though, interesting things are occurring on Haven's moons. The initial heating of a small moon during formation could actually boil off the surface ocean. The gravitational attraction of the moon wouldn't be enough to hold onto the molecules, and they would be lost to space. Such a moon might evolve into something like Jupiter's moon Io, with volcanoes driven by heat flowing out from the interior.

Other moons might be more like Europa. After an initial cooling-off period, they would have oceans—perhaps hundreds of miles thick—covered by a relatively thin layer of ice. Tidal heating, together with heat from the interior, keeps these oceans liquid, and perhaps even keeps them at temperatures above that of Earth's surface. Such moons, like Europa, are prime candidates for the development of life.

But it is the farther, Earth-sized moon that interests us most. Let's call this moon Haven-5, following our assumption that Haven, like Jupiter, has four Galilean-type moons. Because of its size, the heat from Haven-5's initial formation would be slow in dissipating. We expect that the initial melting would allow heavy materials such as iron and nickel to sink to the center, so that this moon, like Earth, would have a heavy core. In addition, because of the moon's size, we expect the heating from radioactive nuclei to be intense. Consequently, we expect this moon to exhibit the kind of tectonic activity we saw on 55 *Cancri* e in the preceding chapter. Molten material from the interior would be continuously brought to the surface and perhaps drive its own version of plate tectonics.

The material brought up from the interior would contain gases such as steam and carbon dioxide that would be released at the surface. Unlike moons such as Io, however, Haven-5 is big enough to hang on to those gases, so it would acquire an atmosphere, probably an atmosphere rich in carbon dioxide and maybe molecular hydrogen. And this leads us to the most surprising feature of this world—the existence of something very similar to a greenhouse effect.

We are used to thinking about Earth's greenhouse effect this way: sunlight comes to the surface because the atmosphere is largely transparent to visible light. The energy, however, is sent back into space as infrared radiation, which tends to be absorbed by carbon dioxide. Thus, the presence of carbon dioxide raises the temperature of the surface above what it normally would be. The key point to notice in this scenario is that the source of heat coming to the surface isn't what's important—the warming effect operates when that heat is radiated upward toward space into an atmosphere that contains carbon dioxide.

Haven-5 would have no incoming starlight, but it would have sources of intense heat in its interior, and this heat would cause its surface to radiate in the infrared. As this radiation tries to escape into space, it would be absorbed by carbon dioxide in the atmosphere, and the surface of the moon would be warmed just as Earth's surface is warmed by the greenhouse effect. We can even imagine scenarios in which the combination of internal heat and atmospheric warming keeps the surface temperature high enough to maintain liquid oceans. Haven-5, in other words, would look just like Earth, except that there would be no sun in the sky. To beings like us, who interact with the world by using visible light, the moon's surface would be eternally dark—the darkness of interstellar space. Any life that evolved on the moon would most likely have "eyes" that saw infrared radiation, since there is abundant heat on

Haven-5. (We note that some animals on Earth, such as pit vipers, have organs that "see" in the infrared.)

So now it is clear why we called our imaginary rogue planet Haven. Despite the absence of a star, its moons could have liquid oceans, perhaps even surface oceans. In the cold depths of interstellar space, its moons could indeed be havens for life, possibly even life on the surface.

However, we have to recognize that our chances of detecting a system like Haven are not very good. It won't shine by reflected light, since it is near no star and it has no light source of its own. Our telescopes won't pick it up, and it would have to pass improbably close to Earth for us to pick up its signature in the infrared. But such habitable rogue planets might be ubiquitous.

Haven—immortal but invisible.

Caveats

Obviously, the system we've described is fictional. We argue, however, that if rogue planets are really as common as estimates suggest, then there is almost certainly a "Haven" out there waiting to be discovered.

The existence of a surface ocean on a large moon isn't impossible, but its existence would depend critically on the amount of internal heat flowing to the surface. With little tidal heating, this heat would have to come from the heat of formation and radioactivity. Both of these sources diminish with time, so eventually the moon's surface oceans would freeze over. The insulating properties of the ice would probably be good enough to keep water beneath it liquid for a long time.

The question of whether life could develop in such an environment is one on which reasonable people differ. We'll discuss the origin of life on Earth in chapter 11, but here we simply note that life on our own planet developed very quickly once it was able to do so. It is conceivable, therefore, that life on Haven-5 could have developed while the moon had a surface ocean, or around deep-sea vents if it didn't. We'll leave the question of what that life might be to science fiction writers.

8

ICE WORLD

The ice was here, the ice was there,
The ice was all around.

Samuel Taylor Coleridge,
"Rime of the Ancient Mariner," 1834

It's cold.

Really cold.

We don't mean Minnesota-in-the-winter cold. We mean cold within shouting distance of absolute zero.

Off in the distance is a range of mountains as tall as the Rockies. They're made of water ice—hard as steel at these temperatures. There's a thin layer of reddish material on the ground: carbon compounds. The sky is black, changing to light blue on the horizon. In the distance, the planet's star is a marble in the sky, scarcely distinguishable from other stars.

This is Ice World.

Most planets in the galaxy that orbit stars are far away from their stars. The energy that reaches these planets—such as Pluto and other Kuiper belt objects (KBOs) in our own solar system—from their stars is negligible. Whatever happens on these planets has to be driven by some other type of energy.

In what can only be considered another example of surface chauvinism, it has traditionally been assumed that such outer planets had to be dead worlds—no tectonic activity, no climate variation, and certainly no life. Yet it is important to remember that these planets are numerous. In our own solar system, for example, there may be as many as 100,000 of these objects in the distant regions beyond Neptune, and there's no reason other systems wouldn't have a similar number. Thus, Pluto may well serve as a good model for what we'll find out there.

The flyby of Pluto by the *New Horizons* spacecraft in 2015 has given us a completely new picture of what these outer worlds are like. (Truth in advertising: one of the authors—MS—is a co-investigator on the *New Horizons* mission.) One of the first surprises in the data coming back from Pluto was the sheer complexity of the planetary surface. Because the surface temperature is only 40 degrees Kelvin above absolute zero (about −230°C or −390°F), familiar materials assume unfamiliar forms. We see clear evidence, for example, of mountains made of water ice that stand about 3 kilometers (2 miles) above the surrounding terrain and most likely extend a similar distance beneath the surface. (For reference, the Rocky Mountains are, on average, about 3 kilometers high.) That mountains this high could be made from water ice, rather than rock, may seem strange, but at these temperatures, ice made from pure water has the tensile strength of steel (although the presence of dissolved impurities or granulation can greatly reduce this strength).

Some of these mountains seem to be made from blocks that formed when a thick ice sheet broke up, a fact that indicates that, far from being "dead," something similar to tectonic activity has occurred on the surface of Pluto in geologically recent times. *Something* had to move that initial ice sheet to break it into the blocks we see today.

Other puzzles abound. Some regions are clearly glaciers, albeit probably made of nitrogen, with nearby dried-up "lakes" of glacial melt. These old lakebeds, as well as much of the glacier surface itself, have no impact craters on them, which means that they are young, geologically speaking—probably no more than 10 million years old.

If you were to stand on the surface of Pluto, you'd see the sky as iridescent blue. So apparently Earth-like, the color comes from a different process than that operating on our home planet. On Earth, the blue sky results from the scattering of sunlight off molecules of oxygen and nitrogen in the atmosphere. On Pluto, it seems to result from a global atmospheric haze with numerous imbedded layers. A Los Angeles–type "smog" could be scattering the sparse sunlight, but if this is true, then there are major differences between the smog on Pluto and the smog in L.A. For example, Pluto's haze occurs where the atmospheric temperature is at its maximum—not at the minimum, as in terrestrial smog regions—and at the moment the best explanation for the layers is that they are produced by condensation of hydrocarbons in the temperature "troughs" of vertically propagating buoyancy waves.

Yet the most interesting thing about Pluto isn't any particular surface detail, but the fact that there had to be a force acting to produce all these surface features—something analogous to the mantle convection that moves continents around on Earth. And regardless of what that something might be—a topic to which we'll return in a minute—one astonishing fact emerges from our calculations: that unknown force must, in order to have generated the surface we now see, produce high enough temperatures in Pluto's interior that materials, such as the ice mountains mentioned above, can be moved around. One interesting implication of all of this is that the temperature might be high enough to melt much of Pluto's interior. And, as it happens, much of that interior is water, so that our current theories tell us that Pluto, like the moons of Jupiter and Saturn, actually has a subsurface ocean.

Water on Pluto, so far from the Sun—who would have thought it? Yet if you step back for a moment and look at the historical record of the discovery of water in the solar system, it makes a kind of sense. In earlier days of astronomy, when surface chauvinism ran rampant, it seemed pretty clear to everyone that liquid water would be found only on our home planet. In fact, one of the authors (JT) once waxed eloquent on the notion that the absence of water in places beyond Earth would be the ultimate limit on our ability to move into space. Then things began to change. It started with the discovery of the subsurface ocean on Europa—a discovery that should have cured us of surface chauvinism but for some reason didn't. This was followed quickly by the discovery of subsurface oceans on other moons of Jupiter and Saturn. In one case (Enceladus), researchers were even able to fly a spacecraft through a geyser and detect evidence of the existence of geological activity on a moon that had been thought dead.

After the initial shock wore off, scientists realized that there was a simple mechanism—the tidal heating discussed earlier—that could account for the unexpected liquid water on these moons. Once this was understood, scientists assumed that subsurface oceans would be found on the moons of large planets. But if the existence of an unexpected underground ocean on Pluto is verified, it would lead to another startling conclusion. If many other KBOs have oceans as well (and the larger ones might), it could well be that most liquid water in the solar system lies in its outer reaches, *not* on the conventional planets.

To support such an unusual suggestion, we would have to identify a mechanism responsible for keeping the subsurface ocean on Pluto from freezing. It can't be tidal heating—there are no large planets nearby. It can't be energy from the Sun—it's too far away. It probably can't be radioactivity, either—Pluto is too small to pack in enough radioactivity to supply the necessary heat, and if the KBOs have the same composition as the rest of the solar system, the same is true of them.

So what could it be? The short answer is that we don't know. There are, however, a couple of ideas floating around.

One possibility that might resolve the difficulty is the phenomenon of phase change. The phase changes most familiar to us are the freezing of water into ice and the melting of ice into water. You probably don't think of the process by which your refrigerator makes ice cubes as a heat source, but the fact of the matter is that it is. The heat given up by the water as it freezes has to be taken away, and that is the job of the refrigerator. In fact, some of the heat you feel in the air that comes off the back of the machine is contributed by the freezing water. (Just put your hand down near the floor when you hear your refrigerator running—you'll feel the heat.) An analogous process occurs when steam condenses into liquid water. As the liquid forms, a lot of heat is given off. This explains, incidentally, why burns involving live steam are so serious—the extra heat given off in the phase change adds significantly to the injury caused by the hot water.

Phase changes occur whenever the atoms and molecules of a substance rearrange themselves in response to a change in temperature. The freezing of water into ice is one example of a phase change, but there are many others. We discussed another in chapter 7 when we talked about helium condensing into a liquid and falling as "rain" in the atmospheres of the Jovian planets of our own solar system. In the cases of Jupiter and Saturn, the phase change is responsible for forming helium droplets that fall inward toward the planet's interior. That converts gravitational potential energy to thermal energy, and that thermal energy then is carried back to the surface by convection. In the case of a planet such as Pluto, there might not be any other sources of energy, so the phase change process would have to carry the entire burden. This creates a problem, because if we don't know what materials might be involved in the phase change and what the properties of the process might be, we can't make realistic calculations of how much heat it could deliver.

Another suggested source of energy (although, to our minds, a more speculative one) is radioactive heating. We know that, ultimately, all elements heavier than iron are made in supernovae and blown back into the interstellar medium to become the building material for new solar systems. You can imagine a scenario in which a few nearby supernovae went off just as our solar system was forming. You can also imagine that the brunt of the absorption of the new material was borne by the gas and dust in the outer regions of our nascent solar system. In that case, objects that ultimately condensed from this material might have a higher concentration of radioactive nuclei within them, and thus a higher level of radioactive heating. (For reference, about half of the energy that heats Earth's interior comes from the decay of radioactive nuclei.)

When we consider these kinds of speculations, we have to keep in mind that how the problem of Pluto's energy source is resolved has deep implications for what we will find in the rest of the galaxy. For one thing, it is implicit in our argument that Pluto is typical of other KBOs. There's no reason to expect this statement to be false, but, on the other hand, we have no direct evidence to support it, either. Charon, Pluto's largest moon, also shows evidence of recent (geologically speaking) resurfacing. This suggests that at least one other object in the outer solar system may have an unidentified interior heat source. But Pluto and Charon are currently thought to have been formed by a collision with another KBO. That they are similar may be no big surprise. But what about other KBOs? Most of them are smaller than Pluto and Charon, and we don't know whether their surfaces suggest additional heating. The *New Horizons* spacecraft is on its way toward an encounter with a much smaller KBO, an encounter that is planned to occur in January 2019. This object, 2014MU69, may be too small to have surface geology.

Then there is the question of whether other solar systems have structures like our Kuiper belt. Again, we have no reason to suppose that they don't, but collections of small bodies orbiting far from distant stars are difficult to detect. We have, however, seen evidence of large amounts of water in the outer regions of newly forming planetary systems, so regions such as the Kuiper belt may not be rare.

Finally, even assuming that Kuiper belts are common, there is the question of Pluto's heat source. If it requires a serendipitous collection of supernovae to seed the distant planets in a system, then worlds such as Pluto could be rare. If, however, Pluto's heat is generated by a routine process such as phase change, then it is entirely possible that every exoplanet system has its own quota of Ice Worlds, complete with subsurface oceans.

What about Life?

As we have stressed repeatedly, the interest in finding liquid water in the galaxy is tied to the notion that liquid water is necessary for the evolution of life. This, after all, is what has driven speculation about the Goldilocks planet. And if we are right to think of Pluto as being typical of planets on which we will find subsurface liquid water, then we should consider the possibility that ice planets may be the home of other living systems in the galaxy.

At first, the idea of life developing in total darkness hundreds of miles below a planetary surface may seem strange. But we know that life exists on Earth far underwater—in fact, complex ecosystems cluster around deep-sea vents, obtaining their energy from chemicals brought to the surface by tectonic activity. In these ecosystems, there are many multicelled organisms, which indicates a long evolutionary history. In fact, as we shall see in chapter 11, an argument can be made that terrestrial life actually began in those deep-ocean vents and only later migrated to the surface.

Since tectonic activity seems to exist, or has existed in the recent past, on Pluto (and presumably on the other larger KBOs), it is not impossible to imagine that similar life forms might develop in subsurface oceans far from the warmth of stars. In fact, just as these ice planets might hold most of the liquid water in the galaxy, they might also be the home of most of its life.

This conjecture is supported by the red organic dust we see on Pluto's surface (although the term *organic*, when applied to molecules, simply means that there are carbon atoms present in their structure—it does *not* mean that the molecules came from living systems). Nevertheless, the presence of these molecules indicates that complex chemical reactions have taken place on Pluto in the past and are probably going on today. And in the end, what is a living system but a series of complex chemical reactions?

On Earth, the advent of what we would call complex, intelligent life took place after oxygen accumulated in the atmosphere and became available as a source of metabolic energy. Oxygen, in turn, was a byproduct of photosynthesis—a surface effect. Whether such a high-grade energy source could appear in a subsurface ocean is not known. It could be that only the largest of the existing ice planets can have sufficient energy to support complex life. If they can do so, however, then we can imagine a complex technological society developing in such an ocean of liquid water, just as our own developed in an "ocean" of gases. Of course, as we shall discuss in chapter 13, there are many other steps involved in going from multicelled life to a technological civilization beyond having lots of energy available.

As we pointed out in chapter 2, such a civilization would be unlikely to develop astronomy as one of the earliest of its sciences, as happened with humans. We can, however, imagine intrepid explorers eventually drilling through layers of overlying ice in their urge to explore the universe beyond their home. What would it be like for them to see the stars for the first time?

What a great subject for science fiction!

Caveats

In a sense, this entire chapter is a series of caveats based on our first close-up look at Pluto. Here are some of the most important assumptions we have made:

- *Pluto really has a subsurface liquid water ocean.* This idea derives from theoretical calculations, but the evidence of recent geological activity is solid. So some energy source must either be present or have been present recently to drive the geology, and the material through which it acts (or acted) must be in some sense a liquid. The evidence may ultimately implicate a liquid other than water. Finally, the theories are based on preliminary interpretation of the data sent back by

New Horizons. They could be proved wrong by new theoretical calculations and/or new data.

- *Pluto is a typical Kuiper belt object.* Pluto is the largest known KBO, although theoretical calculations suggest that at least one larger object may be discovered. Most KBOs are certainly smaller. Size probably makes a substantial difference in how much energy is generated internally by whatever mechanism. When *New Horizons* flies by 2014MU69 in 2019, we will have a better understanding of what smaller KBOs are like.

- *Other planetary systems have Kuiper belts.* We now know that the interstellar medium, especially the giant molecular clouds where new stars are formed, has abundant water. We also have seen water in the outer regions of at least one star and planet system currently forming. So there is no reason to doubt that Kuiper belts are common, but there's little evidence to support it, either. If it's true, however, it would mean that most of the planets circling stars in the galaxy are Ice Worlds like Pluto.

- *Life can develop in subsurface oceans.* Actually, this is probably the least uncertain statement in our list of caveats. Whether it started there or not, we know that multicelled life can thrive in deep-ocean vents on Earth. Many scientists, in fact, argue that the deep-ocean environment, sheltered from the Sun's ultraviolet radiation, is an ideal place for life to start.

KEPLER 186F

SUPER EARTH, ARCHIPELAGO WORLD

It is often said that all the conditions for the first production of a living organism are now present . . . in some warm little pond.

Charles Darwin, letter to J. D. Hooker, 1871

Near the horizon, a small sun sinks, casting a reddish glow over the landscape. The land is low and flat, its lagoons and pools interspersed with muddy banks. Behind us flattened plants stand, their black leaves capturing what energy they can from the faint sun. The breeze seems to move the heavy atmosphere past us, and we seem heavy, too. Welcome to Archipelago World.

o planet is born without a star, but the kind of star a planet forms around exerts a powerful influence on what happens to it after birth. As we saw in chapter 2, stars and planetary systems are born in the gravity-driven collapse of giant interstellar clouds of gas and dust. In fact, one way of thinking about the lifetime of a star is to imagine it as the deployment of successive strategies to combat the relentless inward pull of gravity.

As we pointed out in chapter 2, the first strategy involves the initiation of nuclear fusion reactions in the star's core, using hydrogen as the basic fuel. These reactions generate an outward pressure that balances gravity as long as the hydrogen in the core lasts. It is, in fact, this outward streaming of energy that makes a star shine. Once the core hydrogen is used up, a number of other, more complex strategies are called on until the eventual death of the star. Depending on its mass, this end state may be a supernova—the type of exploding star discussed in chapter 3—or some other structure, such as a white dwarf or a black hole. In any case, the key point is that every star has only so much fuel to consume, and therefore there is only so much time for its planets to evolve.

You might expect that the bigger a star is, the more fuel it has, and hence the longer its lifetime will be. This turns out to be wrong, however, because the inward pull of gravity is greater for big stars, and hence they have to burn their fuel faster. Consequently, bigger stars actually have shorter lifetimes than smaller ones.

To get some sense of the wide variability of stellar lifetimes, let's shrink the 13.8-billion-year lifetime of the universe down to a single year. In this compressed time scale, the Sun and our solar system formed around Labor Day, and our Sun, changed into a white dwarf, will die around the next Tax Day—April 15.

Very large stars, on the other hand, would live only about half an hour before exploding into supernovae, while very small stars, which can easily have lifetimes in the tens or even hundreds of billions of years, could go on shining for a dozen of our compressed years. Thus, small stars that were made in the early days of the universe will certainly have been shining for much longer than our Sun.

Astronomers have a classification scheme for stars that puts them into categories labeled O, B, A, F, G, K, and M, with O-type stars being the most massive, brightest, and shortest lived and M-type stars being the smallest and dimmest. (The standard Astronomy 101 mnemonic for remembering this sequence is "Oh, be a fine girl/guy—kiss me.") Our Sun is a fairly typical G-type star, with a total lifetime of about 9.5 billion years.

Astronomers often refer to late K- and M-type stars as "dwarves," and, as you might expect, they are the hardest type of star to find. On the other hand, our models tell us that they are probably the most abundant type of star in the Milky Way—by some estimates, they make up as much as 85 percent of the stellar population. Thus, we would expect that most of the planets circling stars in the galaxy are to be found in orbit around dwarf stars.

However, as data from the Kepler spacecraft continue to be analyzed, a new kind of planet—called a super Earth—seems to be emerging as a common type. Super Earths are planets, probably rocky, most with two to three times the mass of Earth, but some with up to 10 times Earth's mass. You can think of them as being somewhere between a large Earth and a mini-Neptune in size. We don't have any planets like this in our own solar system, so in this respect at least, our system may be somewhat atypical.

All of which brings us to Kepler 186f, a super Earth circling an M-type star about 490 light-years away—a common type of planet in orbit around a common type of star. But it wasn't the ordinary nature of Kepler 186f that triggered excitement in the scientific community when its discovery was announced in 2014. Instead it was the fact that Kepler 186f was the first planet found that fit into the category nicknamed the Goldilocks planets.

As we explained in chapter 1, the biological sciences labor under what we call the curse of the single example. We know about only one type of life: life that is like our own. It is based on the carbon chemistry of large molecules interacting in water. On Earth, this means that life originated and flourished in the oceans, and this, in turn, means that the surface temperature of our planet had to remain between the freezing and boiling points of water for billions of years.

Keeping the surface temperature in this range requires a delicate balance. For one thing, stars get slowly brighter as they age, which means that the energy the Sun delivers to Earth increases over time. For another, the composition of Earth's atmosphere changes over time as well—think of the introduction of oxygen by living organisms a couple of billion years ago as an example. The fact that the surface temperature stayed within the necessary range through all of this, except for a few episodes of surface freezing collectively called Snowball Earth, is little short of miraculous.

Scientists quantify this idea by defining something called a continuously habitable zone (CHZ), a region around a star in which it is possible for surface oceans to remain liquid over long periods of time. The location of this zone changes from star to star, since it is farther out for big stars and closer in for small ones. In our system, Earth is the only planet in the Sun's CHZ. Were Earth closer to the Sun, the oceans would boil, and were it farther away, they would freeze solid. (Some recent calculations suggest that Earth is actually near the inner edge of the Sun's CHZ, a fact that, if true, presages a hot future for our planet as the Sun warms and the CHZ moves outward. We might wind up looking like Venus.)

The CHZ concept, incidentally, explains the term *Goldilocks planet*, because, like the porridge sampled by the fictional Goldilocks, the planet has to be not too hot and not too cold but *juuust* right. The concentration of scientific attention on finding life that is "like us," then, comes down to finding an Earth-sized planet orbiting in the CHZ of its star. Kepler 186f was the first such planet found.

Having said this, in our view, the present attention to finding planets in a star's CHZ is misguided. We return to this point more fully in chapter 13, but for the moment we reiterate that most of the water in our own solar system (and presumably in others) is not in surface oceans but in subsurface water on worlds such as Europa. Thus, the current focus on finding a Goldilocks planet amounts to a search for the least likely location of liquid water and, presumably, of life.

Returning to our discussion of Kepler 186f, here are some facts: the planet orbits an M-type star that is only about 4 percent as luminous as the Sun. The planet completes an orbit in 130 days. Because of the faintness of the star, its CHZ is close in, and Kepler 186f is about as far from its star as Mercury is from ours. There are other planets in the system, but they are not in the CHZ. From the amount of starlight that Kepler 186f blocks during transit, we can estimate that its radius is about 10 percent bigger than that of Earth.

Unfortunately, because Kepler 186f is 490 light-years away from Earth, it is extremely difficult to go any further in determining its physical properties, even with instruments such as the Hubble Space Telescope. As we improve the use of transits to determine atmospheric composition—a tactic undergoing intense development right now—the environment on Kepler 186f will become much better understood. But currently, we have to invent some reasonable hypotheses to imagine what a visit to Kepler 186f might be like.

We can do a little arithmetic to explain why making reasonable hypotheses is a good way to proceed. There are about 300 billion stars in the Milky Way, and if the estimate of the abundance of dwarf stars given above is accurate, this means that there are about 250 billion M-type stars. Suppose one in 10 of these has planets, and suppose that one in 10 of the planetary systems has super Earths. This means that there are roughly 2.5 billion super Earths orbiting these stars. If even one in 100 of these planets is in its star's CHZ, there are about 25 million Goldilocks planets out there—planets similar to Kepler 186f.

What this means is that even if one of the assumptions we make about Kepler 186f is wrong, there is another planet in that 25 million for which it is right. We can proceed, then, secure in the knowledge that if our guesses are wrong for this particular planet, they are true for some other of the super Earths that we know are out there.

If we assume that Kepler 186f has a rocky composition similar to that of Earth, we can infer that its mass is about 1.4 times greater than that of our own planet. Thus, like the moon we called Haven-5 in chapter 7, it holds on to its atmosphere. (For reference, someone who weighs about 91 kilograms, or 200 pounds, on Earth would weigh about 104 kilograms, or 230 pounds, on Kepler 186f.) Theoretical calculations also suggest that rocky planets up to several times the mass of Earth exhibit the same sort of tectonic activity that we see on our own planet, so we could expect that the atmosphere of Kepler 186f would probably contain carbon dioxide, nitrogen, and water, at least before life arose on the planet and began changing its atmosphere, as life did on Earth by generating oxygen. Like Earth, Kepler 186f could well have a greenhouse effect because of its atmospheric carbon dioxide, or perhaps some other greenhouse molecule. On the other hand, dwarf stars emit significant amounts of ultraviolet radiation early in their lives, so light gases such as hydrogen and helium probably would have been driven off long ago.

The gravitational force on the planet's surface compresses the gases in the atmosphere, so that the atmosphere is denser than that of Earth. Winds on Earth are generated by a combination of solar heating and planetary rotation. So, if the planet rotates as Earth does, the winds would be at least as strong as they are on Earth. The denser atmosphere, in turn, means that wind-driven erosion of the planet's land masses is much more rapid than it is on Earth. In fact, we would expect mountain

1 Exoplanets come in a wide range of masses, compositions, temperatures, densities, and distances from their central star. There is a continuous spectrum of masses, from planets the size of Mercury to those over 10 times the mass of Jupiter. Compositions range from that of hydrogen to that of iron, and exoplanets can be as hot as molten metal or as profoundly cold as interstellar space.

2 Our solar system, below, shows us three types of planets: terrestrial and Earth-like planets; gas giants; and the recently designated dwarfs (a hypothetical brown dwarf system is at lower right). Telescopic observations and robotic explorations have revealed that each planet is incredibly complex, with a unique story of birth and evolution.

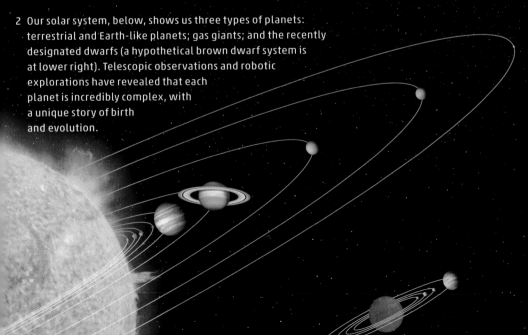

3 Most stars have planets—in fact, observations suggest that on average, each star has at least four. This estimate is derived by extrapolating the frequency of exoplanets that we have observed around stars other than our Sun. Most exoplanets that have been discovered are larger than Earth, so we can expect this extrapolated number to increase as detection thresholds allow us to detect smaller planets. Furthermore, research suggests that most planets are not even bound to stars. This artist's conception suggests the plethora of planetary discoveries made by the NASA Kepler Space Telescope.

5 The region of space that the NASA Kepler Space Telescope searches for planets is
   a tiny cone-shaped area parallel to our local spiral arm of the Milky Way galaxy.
   Kepler monitors the light of only about 100,000 stars—about 0.2 millionths of the
   number of stars in our galaxy—for dips indicative of planetary transits.

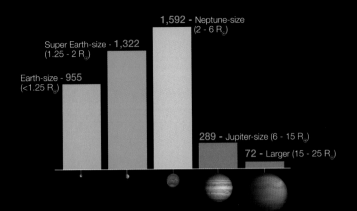

1,592 - Neptune-size (2 - 6 $R_{\oplus}$)

Super Earth-size - 1,322 (1.25 - 2 $R_{\oplus}$)

Earth-size - 955 (<1.25 $R_{\oplus}$)

289 - Jupiter-size (6 - 15 $R_{\oplus}$)

72 - Larger (15 - 25 $R_{\oplus}$)

7 The size distribution of the exoplanets that the Kepler Space Telescope has discovered so far indicates that most planets are about two to three times the size of Earth. This inference is tentative; improvements in the ability of telescopes to detect smaller planets may shift the distribution's peak toward Earth-sized bodies.

8 55 *Cancri* e is in such a close orbit around its central star that its "year" lasts about 17 Earth hours. This planet likely has immensely strong gravitational and stellar wind interactions with its star, so that dramatic electrical "fireworks" might exist in its atmosphere.

9 Rogue planets, untethered from any star, are extremely difficult to detect and characterize. Their diversity might be as broad as that of star-bound exoplanets, but we don't have sufficient information to describe their main statistical properties. There are hints that rogue planets, one of which is illustrated here, may greatly outnumber star-bound planets.

10 Rogue planets in the Jupiter-or-larger size range can retain their internal heat for billions of years. So, other than being dark at visible wavelengths, the evolution of large rogue planets could be similar to that of their star-bound cousins. However, smaller rogue planets might cool off rapidly and quickly become Ice Worlds, like the artist's conception here. How fast that occurs depends primarily on their size.

11 Pluto's large nitrogen glacier, at upper right in this enhanced 2015 image from the NASA probe *New Horizons*, dominates the surface, sitting inside a large depression that might have been produced by an ancient asteroid impact. Such impacts might penetrate deep into Pluto's ice crust and provide a means to mix material from the surface with material (possibly water ice) excavated from the deep interior.

12 Pluto, seen here in an enhanced 2015 image from *New Horizons*, is a type of Ice World, with characteristics that are in some ways similar to those of Jupiter's moon Europa. Its surface is covered by nitrogen, methane, and water ice, and models suggest that the ice might overlie a subsurface ocean that contains liquid water. The composition of Pluto's large dark areas is unknown, but they could be tholin-like material, produced by methane chemistry in the atmosphere, that subsequently fell to the surface and accumulated over Pluto's long history.

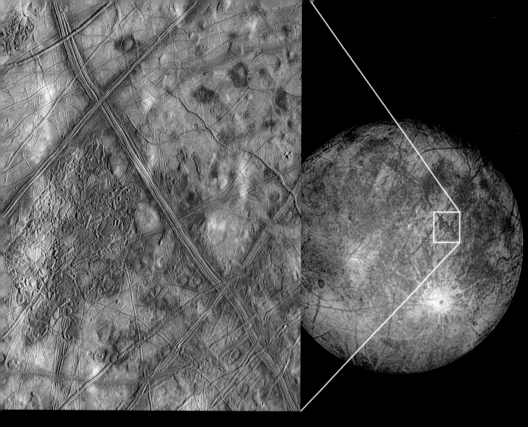

13 The ice crust of Europa, a moon of Jupiter, at above left in close-up, is covered in fissures that appear darker than the surrounding ice. The composition of these darker cracks is unknown, but researchers suspect that they are material that has been forced upward from the ocean below, where it was formed. This darker material could contain signatures of any biological activity that might occur in Europa's ocean.

14 Ice Worlds may be one of the most common types of planets in our galaxy. We have at least six examples in our solar system: Europa, seen below, may have biological significance. Its subsurface ocean is heated by internal tidal dissipation due to gravitational interactions with Jupiter and Io and may have hydrothermal vents and other similarities to Earth's deep-ocean regions.

15 Kepler 186f, seen here in an artist's conception, is one of the most Earth-like exoplanets yet discovered. Its size and mass are similar to those of Earth, so it may have a similar makeup of silicates and metal (nickel and iron) as well. If the planet is not tidally locked, its rotation would produce a magnetic field that shields its surface from stellar ultraviolet light as well as stellar wind—both key features of habitable worlds like Earth.

16 Kepler 186f, like Earth, is located within the so-called habitable zone around its central star, which means that liquid water should be stable on its surface (b, c, d, and e are the other four planets that orbit the red dwarf Kepler 186; not to scale).

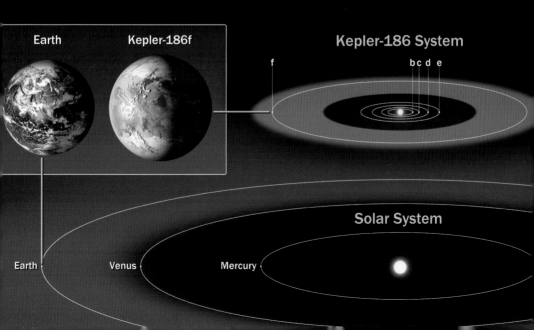

Earth Kepler-186f

Kepler-186 System

f

b c d e

Solar System

Earth   Venus   Mercury

17 Water Worlds have a composition that is mostly H_2O, but they also have other, heavier materials such as metals and silicates in their interiors. That makes them somewhat similar to gas giants but with water replacing hydrogen. Exoplanets' water can exist in a wide range of phases, ranging from steam in the atmosphere to high-density, and per-haps electrically conductive, phases in the deep interior.

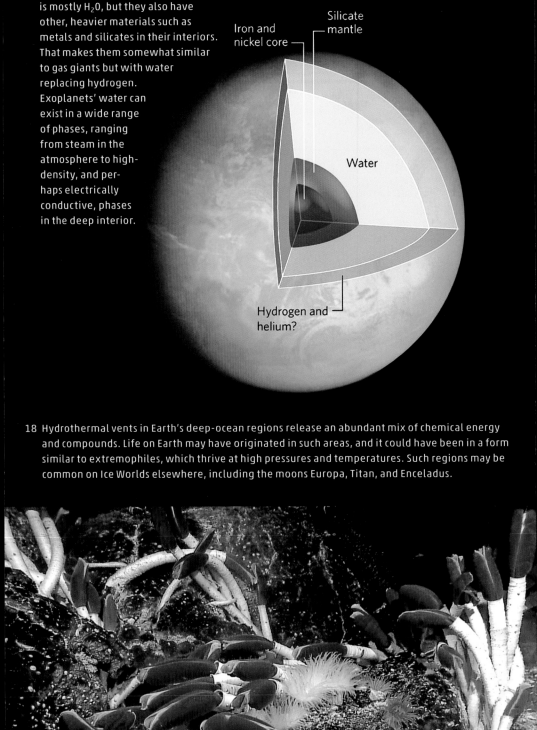

Iron and nickel core

Silicate mantle

Water

Hydrogen and helium?

18 Hydrothermal vents in Earth's deep-ocean regions release an abundant mix of chemical energy and compounds. Life on Earth may have originated in such areas, and it could have been in a form similar to extremophiles, which thrive at high pressures and temperatures. Such regions may be common on Ice Worlds elsewhere, including the moons Europa, Titan, and Enceladus.

19 Once the Late Heavy Bombardment had ended and Earth cooled sufficiently for liquid water to form the first seas, it was only a few hundred million years before single-celled life was ubiquitous on our planet. Earth's surface held a rich "primordial soup" of organic materials from which life may have originated, but evidence that could reveal the details of *how* that life originated has not been found.

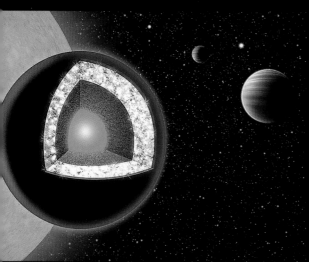

20 We have found exoplanets that appear to be made mostly of carbon, which at high pressure exists in crystalline form. For example, the so-called Diamond World, 55 *Cancri* e, has a surface made primarily of graphite over a thick diamond layer. Such exoplanets are larger than Earth and closer to their host star, so they would be characterized by high-energy fluxes at their surfaces—which could drive the emergence of complex carbon structures, perhaps more complex than those that exist on Earth.

21 Water Worlds can have a mix of water and hydrogen gas in their outer atmospheres. Water vapor is a strong greenhouse gas, so the outer atmosphere might be hot and steamy for those Water Worlds in the habitable zones of their central stars. Surface-atmosphere water on Water Worlds that form in or move to regions far outside of their habitable zones may freeze, creating an ice crust like that seen on Ice Worlds.

22 The universe appears to be rich in the fundamental prerequisites for life: usable energy, carbon and other raw materials, and liquid water. Their abundance on or inside the diverse exoplanets that we have discovered suggests that life could have arisen on numerous planets in our galaxy alone. One technique in the search for life elsewhere in the universe is to examine the chemical biomarkers that can be observed in exoplanets' spectra.

23 Gases that suggest possible biological origin can be detected in the spectra of distant stars as planets pass in front of them. The presence of methane, oxygen, carbon dioxide, ozone, and ammonia, as well as many other gases that are observed in a disequilibrium mix, might imply a continuous release of some compounds, perhaps from biology.

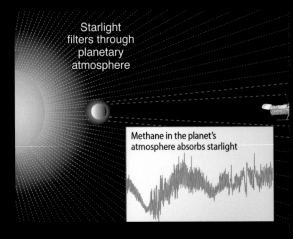

Starlight filters through planetary atmosphere

Methane in the planet's atmosphere absorbs starlight

24 The diverse characteristics of the planets in our solar system make the precise definition of a planet difficult and controversial. The International Astronomical Union's (IAU) official definition of 2006 focuses more on a planet's location than on its intrinsic characteristics and so has confused the issue. For example, the IAU definition would exclude escaped planets such as rogue worlds. The continuum of objects from asteroids to giant planets may mean that a simple definition is impossible.

Moon   Kepler-37b   Mercur

25 NASA's *Curiosity* rover (shown here in a 2015 self-portrait) undertakes its in situ search for life on Mars via direct chemical tests and spectroscopic study of surface materials. Detection of metabolic by-products, such as methane, released by subsurface life is an indirect but powerful means of determining whether life is present. Methane, coming in brief bursts from the planet's interior, has been detected by *Curiosity*, but it is not yet known if it is the result of biology.

ars   Kepler-37c   Earth   Kepler-37d

26 Europa's subsurface ocean presents a major challenge for life detection. Biomarkers of life might be carried upward and onto the moon's icy surface by oceanic material that oozes out of fissures, as seen in this illustration. But chemical processing along the way, as well as radiation from the Jovian environment, may alter their chemical form. Direct detection will require a means of drilling into the deep ocean so that robotic vehicles can move downward to search for life.

27 The so-called water hole is the radio frequency region near the emission lines of atomic hydrogen (H) at 1,420 MHz (a wavelength of 21 cm) and hydroxyl (OH) at 1,666 MHz (a wavelength of 18 cm). If another civilization's biology requires water, as terrestrial life does, this part of the spectrum is an obvious place to look for signals that that civilization might send outward.

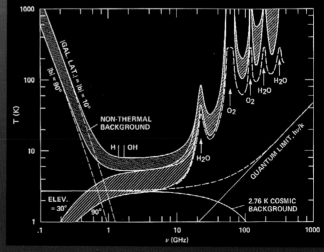

28 A typical galaxy, such as the Whirlpool Galaxy, shown here in a 2005 Hubble image, has about 400 billion stars. The Fermi paradox, named for physicist Enrico Fermi, suggests that if interstellar colonization can occur in 10,000-year steps, our own Milky Way galaxy could be colonized in fewer than 10 steps, i.e., a much shorter time than its 13.7-billion-year history. This raises a question first asked by Fermi himself: "Where is everybody?"

29 The classic SETI technique is to search for electromagnetic signals of clearly non-natural origin. This technique requires extremely sensitive radio telescopes, such as this New Mexico array, because signals from a distant civilization at our level of technological development would be very weak.

30 In September 2016, astronomers using NASA's Hubble Space Telescope
   confirmed the existence of an exoplanet that orbits two red-dwarf stars in
   the system OGLE-2007-BLG-349, eight thousand light-years away from Earth
   toward the center of the Milky Way. This is the first time such a three-body
   system has been confirmed using the gravitational microlensing technique.

chains thrown up by tectonic activity to be worn down quickly, their rocky structure converted into beach sand and delta mud as it is on Earth.

All of which brings us to the question of water. There is a long-standing scientific controversy about how Earth acquired the water that now fills our ocean basins. The current theoretical front-runner suggests that most of this water was brought in by comets and asteroids in what is called the Late Heavy Bombardment, toward the end of that game of cosmic billiards we discussed in chapter 2.

So let's assume that, like Earth, Kepler 186f has acquired enough water to fill some ocean basins but not enough to produce the sort of water world we will visit in the next chapter. The rapid erosion of land masses driven by the dense atmosphere would then result in large areas of low-lying, swampy land, with many open pools and tidal basins. (The easiest way to convince yourself of this is to imagine scraping off the Himalayas and the Tibetan Plateau and dumping them into the Indian Ocean.) Standing on the surface, we would see ponds and islands stretching away in all directions—a landscape that astronomer René Heller has dubbed Archipelago World.

And now we are in a position to understand the most interesting feature of super Earths such as Kepler 186f: they are much better candidates for the development of life than our own planet is.

We will discuss current ideas about the origin of life on Earth in chapter 11, but we point out that Darwin's "warm little pond" remains a viable candidate for the place where the first living cell was assembled. One of the features of Archipelago World is that it is full of such warm little ponds, each of which can be thought of as a separate chemical laboratory, each running its own experiment on the origin of life. This, in turn, means that whatever process led to life on Earth would have many opportunities to be replicated on Archipelago World.

We will also see, in chapter 11, that a popular group of theories for the origin of life on Earth can be labeled "frozen accident" theories. The essential tenet of these theories is that once the basic molecular components of living systems had been produced, the first cell was actually the result of a chance coming-together of just the right molecules to produce life. We know that the more time you can wait for a random event to occur, the more likely you are to see it happen. This is where the long lifetime of dwarf stars becomes important, because, as we have seen, instead of the comparatively short lifetime of a G-type star such as the Sun, dwarf stars keep shining for tens or even hundreds of billions of years. Thus, dwarf stars that formed early in the life of the universe have had a lot more time for life to develop than has Earth. It may well be, in other words, that if we do find life out there (a subject to which we will return in chapter 12), it will be on a planet like Kepler 186f.

We can make one more conjecture about what kind of life might develop on that planet. When one of the authors (JT) lectures on evolution, he often takes on arguments from proponents of what is called intelligent design by asking his students a simple question: why is grass green?

The simple answer is that green light is reflected from grass while other colors are absorbed. But if you think of grass as a solar collector whose job it is to harvest energy from the Sun, this is a stunningly poor example of design. An intelligently designed collector would absorb all the radiation that fell on it and would appear to be black. On a planet such as Kepler 186f, where solar radiation would be much scarcer and more precious than it is on Earth, we would expect that natural selection would drive plants to the most efficient design possible, which is why we talked about black leaves in the introduction to this chapter.

In the end, the most interesting thing about Kepler 186f may well be the fact that it is enough like Earth to allow us to be reasonably confident that we can extrapolate from our knowledge of our own planet to a new world.

Caveats

As we have stressed, most of the above narrative has been driven by assumptions—reasonable ones, but assumptions nonetheless—about the composition and atmosphere of Kepler 186f. For example, to come up with our estimate of the planet's mass, we assumed that it has a composition similar to that of Earth. Other assumptions would produce other estimates—anything from 0.32 Earth masses (in the unlikely event that Kepler 186f is made entirely of water and water ice) to 3.8 Earth masses (in the equally unlikely event that it is made of iron). It is because of this type of uncertainty that we've spent so much time talking about the large number of Kepler 186f–like planets out there.

There are many other uncertainties. We don't know the rate of rotation of the planet and hence can't estimate the length of its day. Given the fact that Kepler 186f is close to its star, it is possible that it is tidally locked, which would mean that it always presents the same face to its star, just as the Moon always presents the same face to Earth. The effects of such a situation on climate and the evolution of life are simply unknown. But in such a situation, we can expect extremely rapid winds from the hot, high-pressure dayside to the cold, low-pressure nightside. We have actually detected such planets among the host of known extrasolar planets, where the tidally locked rotation generates supersonic (up to Mach 10) winds carrying energy from the dayside to the nightside. This could lead to preferred locations, perhaps at the edge of the lighted area, where life would be most likely to exist.

Finally, we recall a common saying in the exoplanet community: "Earth-sized does not mean Earth-like." All we really know about Kepler 186f is that it is roughly Earth-sized, but then, so is Venus, which is about as un-Earth-like as it can be. Furthermore, the fact that a planet is in the CHZ of its star isn't enough to establish the possibility of life—we need to know a lot more details about the planet, such as its atmospheric composition, water content, and so forth. As we learn more about Kepler 186f's atmosphere in the next few years, we will be able to narrow the possibilities for life. It is also likely that we will find other, closer worlds that are more amenable to analysis in the future. So the test of the validity of our assumptions may occur soon.

In the end, however, we can say that if we ever do find life that we recognize out there, it will likely be on a super Earth in the CHZ of a K- or M-type star.

10

GLIESE 1214B

WATER WORLD

And the earth was without form and void; and darkness was upon the face of the deep. And the spirit of God moved upon the face of the waters.

Genesis 1:2

You are on the surface of an ocean. Waves extend out as far as the eye can see into a dark haze. You seem heavier here, and the waves are moving three times faster than they do on Earth. The stars in the sky are dimmer than on Earth but otherwise look ordinary. There is a pleasant warmth from the waters below.

Your instruments tell you that the water extends deep beneath your feet, deeper than even the deepest ocean on Earth. In the depths, the incredible pressure pushes the water into strange configurations— configurations that you've never seen, even in your laboratories. At the very bottom, perhaps 160 kilometers (99 miles) or more beneath the surface, there is solid ground, a planetary surface with mountains and valleys and plains. Residual heat from the planet's formation, along with radioactivity in this solid core, causes its mantle to churn, bringing up hot material in a network of deep-sea vents, and around these vents are complex ecosystems of multicelled life.

This is Water World.

T hink about the oceans that cover 73 percent of Earth's surface. They are in a kind of Goldilocks situation. If there were a lot more water, there might be no dry land, and if there were a lot less water, we might have gone the way of Venus in a runaway greenhouse. It seems that on our planet, the amount of water is just right. Why?

A whole constellation of processes led to our oceans. For starters, when Earth was forming, there was a period when the protoplanet was sweeping up all the planetesimals near its orbit. Someone standing on the surface would have seen a constant infall of massive meteorites, each bringing in energy to be converted into collisional heat. This was a sequence of events that astronomers call the Late Heavy Bombardment. Eventually, the planet melted all the way through and the process we call differentiation started. The heaviest materials—mainly iron and nickel—sank to the center to form Earth's core. Lighter materials such as basalts sank down to form the next layer (what we call the mantle), and the lightest materials, such as granite, bubbled up to the top. As in a salad dressing left out too long, Earth's materials separated.

As it happened, there was enough of the lightest material to cover about a quarter of the planet's surface and become the base for what we call continents. The planetary surface was divided into two parts: the continents, made from lightweight granite, and the deeper ocean basins, made from heavier basalt. The existence of ocean basins, then, depends on two things: the process of differentiation, which is probably a universal factor in the formation of terrestrial planets, and the fact that there was only enough lightweight material to cover a fraction of the planetary surface—a subject to which we return below.

So, differentiation gives us a planet with ocean basins, like bathtubs waiting to be filled with water. And this leads us to the next question: where could that water come from?

As it happens, this is a question about which scientists have argued for decades. There are three possible sources of Earth's water: it could have been locked into the planet, to be released back to the surface by volcanoes in a process called outgassing; it could have been brought in by meteorites during the Late Heavy Bombardment; or it could have been brought in by comets. At the moment, the developing consensus seems to be that all of these sources contributed to Earth's supply of water, and the scientific debate has shifted to discussions of the relative importance of each source and the details of the process—identifying the type of asteroid that was the biggest source of water, for example.

The main tool for resolving this debate depends on the kind of water found in the various possible sources. We're used to thinking of water as good old H_2O—two hydrogen atoms and an atom of oxygen. The reality, though, is a little more complicated than that. On Earth, about 0.15 percent of hydrogen atoms have a nucleus that is not a single proton, as in normal hydrogen, but a proton and a neutron. This version of hydrogen is called deuterium. The water that incorporates such an atom, heavy water, is denoted by the symbol HD. Our theories tell us that the deuterium in the universe was created in the Big Bang and that its abundance hasn't changed much since then. In addition, theoretical calculations tell us that the deuterium on Earth was in the interstellar cloud from which our system formed and was not altered much by processes that took place during the formation of Earth. HD abundance, in other words, becomes a way to evaluate the ability of each source to supply our planet's water. Based on measurements of HD in comets and meteorites, we'll take a ratio of 20:40:40 to be a reasonable estimate of the percentage of water brought in by outgassing, asteroids, and comets, respectively.

This brings us to the process of filling the ocean basins. On Earth, as we pointed out above, enough water was brought in to fill the basins but leave areas of dry land. We cannot, however, assume that this is always the case among exoplanets. In fact, we have already identified two variables that control the configuration of surface oceans: the amount of light material that comes to the surface during differentiation and the amount of water brought in by the mechanisms outlined above.

Neither of these variables can be expected to be uniform over the collection of exoplanets. Even in our own solar system, the distribution of water is far from uniform. Take the moons of Jupiter as an example. Io, the innermost moon, has no water, while Europa, Callisto, and Ganymede all have subsurface oceans.

This means that we have to be prepared to encounter a wide variety of possible worlds with liquid water on their surfaces.

Ocean Worlds

Like Earth, these worlds have only enough light material to form continents that cover part of their surface. They also have only enough water to fill their ocean basins, but not enough to cover the continents. They would look familiar to us, with oceans, beaches, coastlines, and the like.

Water Worlds

There are several ways that a world entirely covered with water could arise. It could, for example, simply be a result of the influx of massive amounts of water—water that would cover any continents that might have formed. It could also arise on a world in which there was so much light material that it covered the entire surface. Water on this world would have no basins to flow into, so it would instead form a worldwide ocean. If there wasn't much water coming in, on the other hand, we could have something like the Archipelago World we visited in the previous chapter.

Having said this, we should note that we know of six exoplanets that might qualify as Water Worlds. These planets are super Earths—intermediate in size between Earth and Neptune. The first one discovered was Gliese 1214b. First detected in 2009, the planet orbits a star about 42 light-years away in the constellation Ophiuchus. (The name of the constellation translates as "snake bearer," and it's usually pictured as a man holding a snake.) A combination of measurements of the planet's atmospheric spectrum and mathematical models causes astronomers to suspect that it might be what we can call an Ocean Planet or Water World.

The planet's name refers to the German astronomer Wilhelm Gliese (1915–93), who spent much of his life producing a catalogue of nearby stars. Gliese 1214, then, is the 1,214th star in that catalogue. Gliese's work was interrupted by World War II: he was conscripted and sent to the Eastern Front, where he spent four years in Soviet captivity before being repatriated.

Steam and Ice Worlds

If we add temperature to our list of variables, we find even more variability in our array of possible worlds. A Water World that has just been hit by a large, Ohio-sized asteroid could easily turn into Steam World, a planet surrounded by an atmosphere of live steam formed by the boiling of its oceans. Theorists have, for example, suggested that Earth may have been this sort of planet for various periods during the Late Heavy Bombardment. This is why we have always referred to living things here as descendants of the last experiment in the origin of life—there is no way to tell how many times life developed, only to be wiped out by a major collision.

We could also get a Steam World in the early stages of development if, for some reason, there was a high concentration of radioactive material in the newly formed planet. This Steam World state would persist until the radioactivity died down and the temperature of the surface fell below the boiling point of water. At this point, rain would start to fall and ocean basins (if there were any) would start to fill up. A Water World very close to its primary star could also be a Steam World due to the greenhouse effect.

Finally, if the planet were far from its star, we would have something like the Ice World we visited in chapter 8. Such a world might have a subsurface ocean, but its surface would be frozen solid. However, even a planet such as Earth can go through periods much like what we found on Ice World. These "Snowball Earths" were periods in our history when the oceans froze over, so that almost all of the planet was covered in snow and ice. Such a planet reflects a lot of incoming sunlight, so it is unlikely that the Sun eventually melted the ice. In fact, we think that it was volcanoes spewing carbon dioxide into the atmosphere and creating a greenhouse effect that brought the planet out of its Snowball phases.

Having examined the possible combinations of surface configurations and water supply on planets with surface oceans, we can go back to Gliese 1214b and see what its globe-girdling ocean is like. To do this, we're going to have to examine the properties of one of the strangest substances nature has ever produced—water.

Water is strange? Yes. Although its ubiquity in our lives makes it seem ordinary to us, water has some very odd properties. For example, when most substances change from a liquid to a solid, they shrink—the volume of the solid is less than that of the liquid. This is what you would expect, since the molecules in a liquid are free to move around while those in a solid are locked into a rigid structure. But water doesn't follow this pattern. Chill liquid water and it shrinks until it reaches 4°C (40°F). Chill it more and it starts to expand, and it will expand even more when it turns to ice.

This fact can cause problems in plumbing systems, where pipes that are allowed to freeze burst as the ice expands. On the other hand, if water didn't have this unusual property, life on Earth would be very different. When a pond freezes in the winter, for example, the ice, being less dense than the water, floats to the top, insulating the water beneath. If this didn't happen—if the ice sank to the bottom—the pond would freeze solid, probably killing its aquatic creatures in the process.

Many other anomalies arise when we look at water. To mention just one, there is a general correlation between the atomic weight of a material's molecules and its boiling point—the higher the weight, the higher the boiling point. If you used this relationship to calculate the expected boiling point of water, however, it would be expected to boil at −93°C (−135°F). The fact that we can swim in liquid water when the temperature is a balmy 27°C (80°F), then, is simply another illustration that water doesn't follow the usual chemical rules.

The only way we can understand the behavior of any molecule is to examine the nature and arrangements of the atoms that make it up, and water is no exception. The familiar H_2O formula hides some interesting facts about the water molecule. The hydrogen atoms arrange themselves around the oxygen molecule like a couple of Mickey Mouse ears, but with an asymmetric 105-degree angle between them. Because of the nature of the interatomic forces, the electrons in the molecule tend to spend most of their time around the oxygen atom. Thus, the water molecule can exert an electric force on other molecules, even though it has no net electrical charge. (This arises because if the negative part of the molecule is closest to an external object, it exerts a larger force on it than the more distant positive part.)

This arrangement of electrons, incidentally, explains why soaking a greasy cooking pot in water overnight helps to clean it up. The electrical charges on the water molecule exert a force on the molecules in the grease, pulling those molecules off one at a time.

The electrical arrangement in water molecules also affects the way these molecules arrange themselves in the liquid state. If you took a snapshot of the molecules in liquid water at any moment, you would find them arranging themselves at the corners of a tetrahedron (that's a pyramid-shaped geometrical figure). Follow a single molecule around, however, and you would find it flitting from one tetrahedron to another; unlike the situation in a normal liquid, water molecules do not move around randomly. Liquid water is not a solid, but it's not quite a normal liquid, either. One of the authors (JT) characterized this property by noting that water never quite forgets that it was once ice. This fact probably explains most of its strange properties.

What will we find, then, when we go beneath the surface of Water World?

At the surface, the water will look normal to us, but as we descend into the depths the pressure will start forcing the water molecules into specific arrangements. Increase the pressure by going deeper and the molecules will be forced into a different arrangement. Each of these shifts is a phase change, analogous to freezing or boiling. We know of at least 13 phases of water, and there could well be more in the deepest oceans on Water World.

The question of whether there might be life on a Water World is one that we probably won't be able to answer for some time. From the planetary densities, we can assume that, like the Jovian planets in our own system, Water Worlds have rocky cores. If those cores exhibit something like plate tectonics, then at the bottom of the oceans, there should be deep-sea vents similar to those on Earth. As we will point out in chapter 11, many theorists argue that life on Earth originated at those vents and moved to land later. The vents on Water World, then, could have had the same sort of history and support the same sorts of ecosystems. Whether that life could develop into something more complex (think fish or swimming dinosaurs) is a matter of conjecture.

Caveats

The existence of a Water World depends on the results of differentiation and the amount of water coming into a planet. Given the enormous diversity we see among exoplanets, we find it hard to believe that the conditions outlined above haven't been met somewhere in the galaxy. We cannot, however, estimate how often Water Worlds occur—that is a question that will be dealt with by future explorations.

We should also emphasize that the identification of Gliese 1214b as a Water World depends on models of planetary structure and could be wrong. We argue, however, that given the number of exoplanets out there, there are sure to be planets for which this identification is correct.

Finally, the existence of life on Water Worlds is very much an open question. If, for some reason, the deep-ocean vents on these planets are different from those on Earth—if, for example, they bring up materials that cannot easily serve as an energy source—life might not develop at all. In addition, if it turns out that the origination of life depends on a surface phenomenon such as the existence of tidal pools, then Water Worlds might well be completely sterile.

11

LIFE ON EARTH

*There is grandeur in this view of
life, . . . [that] . . . from so simple
a beginning endless forms most
beautiful and most wonderful have
been, and are being, evolved.*

Charles Darwin, On the Origin of Species

*For I can trace my ancestry back to a
single primordial protoplasmic globule.*

W. S. Gilbert and Arthur Sullivan, The Mikado

Before we start to think about what sorts of life might
have developed on exoplanets, it is probably a good idea
to understand something about the way life developed
on our own world. It is, after all, the only place where we can be
absolutely sure that life has appeared. Despite the fact that the

life sciences suffer from the curse of the single example, it makes sense to explore how Earth came to be the abode of Darwin's "endless forms" before we go out into the galaxy.

As we pointed out in chapter 1, life on Earth developed in two stages, with the first stage being the appearance of the first living cell from inanimate matter, and the second stage being the proliferation and diversification that led to today's biosphere. These stages are often referred to as chemical evolution and evolution by natural selection, respectively. As it happens, we know a great deal about the second stage and somewhat less about the first. So let's start with what we do know: let's imagine that the first cell has appeared somewhere, and ask how we got from there to here.

Evolution by Natural Selection

We've probably all heard the story of Darwin's voyage on the *Beagle* and his visit to the Galápagos Islands. Like Newton's apple, Darwin's finches have become a standard part of the folklore of science. For our purposes, however, the details of how he came to his ideas is not as important as how those ideas have been expanded and worked out into our modern theory of evolution. In essence, the theory is built on two propositions:

1. There are always differences among members of a given species—differences that can be passed from parent to offspring.
2. There is always competition among members of a given species for whatever goods the environment provides.

Let's look at these propositions separately. The first is simply a statement of modern genetics. If your parents have a gene for a particular characteristic—height or eye color, for example— there is a probability that you will inherit that gene and display that characteristic, too. When Darwin was writing in the mid-nineteenth century, however, Gregor Mendel was just introducing the concept of the gene in what is now the Czech Republic. Thus, although Darwin knew that there was ample evidence to establish the truth of the first proposition, he had no idea why it was true. It wasn't until the 1920s, in fact, that modern genetics was incorporated into evolutionary theory in what has come to be called the Grand Synthesis. Today, of course, we know that DNA mutations produce much of the variation within a given species.

One of the most charming features of Darwin's *On the Origin of Species* is a long discussion of the art of pigeon breeding in the first chapter—you can just picture Darwin hanging out in an English pub with other pigeon fanciers. He uses practices such as cattle and pigeon breeding to introduce the concept of artificial selection: the ability of humans to produce birds or flowers of a certain color, cattle with certain growth patterns, or dogs capable of acquiring specialized skills such as sheepherding or hunting. This process was well known to be governed by the careful choice of mates by breeders, a process that locks in beneficial mutations. You couldn't, after all, breed Angus steers or Labrador retrievers unless genetic properties could be passed down from parents to offspring—unless, that is, the first proposition above was true.

From there, it is a small step to ask a simple question: if human beings can guide the development of species by conscious choice, can nature guide a similar development without conscious choice? And this brings us to the second proposition, which provides the basic mechanism for such a process. The idea (again, in modern language) is that if a particular genetic variation gives an organism an advantage—makes it easier to avoid predators or acquire food, for example—then that variation is more likely to be passed on to future generations. Thus, although no single individual can change the genetic endowment with which it is born, the genetic composition of the population can change over time because of this winnowing process, which is known as natural selection. It was this rather simple and straightforward process that led from the single first cell to the millions of species that inhabit our planet today.

There is massive evidence, both in the fossil record and in modern DNA studies, to back up this picture of the diversification of life. A couple of features of this history need to be stressed, however. First, although life forms do, indeed, become more complex over time, there is no overriding "goal" or "purpose" to evolution. Evolution did *not* take place specifically to produce the human race, as some Victorians thought. The overall impression you get when looking at the history of life is of a process that tries out every blind alley until it finds something that works.

Having made this point, we need to recognize that, if we are going to look for life "like us" on an exoplanet, some conditions will have to be met. First, on whatever planet we are searching, both of the above propositions have to be true. There must be something analogous to our terrestrial genetic system, so that something like genes can be passed down through the generations. In addition, there has to be competition, some reason for changes to confer an advantage. An ideal tropical paradise with abundant food and shelter may produce creatures that are fat and happy, but it won't operate according to the laws of natural selection.

In any case, that is how life developed from that first cell in the one world where we know life exists. Let's turn now to the first and most mysterious question: where did that original cell come from?

Chemical Evolution

Until the middle of the twentieth century, the question of how life could arise from inanimate matter just wasn't something scientists felt was a legitimate field of study. Living systems were known to be complicated, and the general feeling was that the origin of life had to be equally complicated—complicated enough to be well beyond the reach of science at the time. Leave the origin-of-life research to the priests and philosophers, the attitude seemed to be, and work on a problem that can actually be solved.

Then, in 1952, in the basement of the chemistry building at the University of Chicago, that all changed. Harold Urey (1893–1981), who had already received a Nobel Prize, and his graduate student Stanley Miller (1930–2007) did an experiment that completely changed the status of origin-of-life research. The goal of their tabletop experiment, the results of which were published in 1953, was simple: they wanted to build an apparatus that mimicked the early, prebiotic structure of Earth.

There was a glass sphere partially filled with water (to represent the oceans) and heated (to simulate the effect of the Sun). There was an "atmosphere" of methane, water, hydrogen, and ammonia (the best guess people had at the time about the composition of the early atmosphere) and electric sparks to simulate lightning. Urey and Miller turned on the experiment, and after a few weeks the water in the sphere turned a murky reddish brown. Analysis showed that, starting with simple nonorganic molecules, they had produced a class of molecules known as amino acids—molecules that play an important role in living systems.

A word of explanation: when we say that the molecules in living systems are complex, we mean that they are complex in a special way. The main molecules in living systems are modular—that is, they achieve complexity by assembling simple building blocks in complex ways. For example, molecules called proteins play a crucial role in regulating the chemistry of living systems. The building blocks of proteins are relatively simple molecules called amino acids, and proteins achieve complexity by stringing different amino acids in a chain, in much the same way a necklace achieves complexity when you assemble different beads on a string. What Urey and Miller found, therefore, was that the basic building blocks from which the molecules of life are assembled could be produced by simple reactions among inorganic molecules.

We can make a couple of points about the Urey-Miller experiment. First, the consensus today is that the atmosphere of the early Earth probably wasn't like the one used in their original experiment, so the precise process the pair found probably never actually took place. The importance of the experiment is that it moved the origin-of-life question firmly into the realm of science. Second, subsequent experiments of a similar type produced segments of all the molecules found in living systems, up to and including DNA. And finally, none of this discussion matters much because over the last half of the twentieth century, simple organic molecules—the building blocks of life—have been found in meteorites, in comets, and even in interstellar dust clouds. Thus, it appears that these basic building blocks could have come to Earth (or any exoplanet) either through a Urey-Miller type of process or by being introduced by comets or meteorites. The real origin-of-life problem, then, is not how to produce the basic building blocks, but how to assemble those blocks into something we would recognize as a living cell. That's where the last gap in our knowledge needs to be filled in.

There are three questions we can ask about the creation of that first cell: (1) Where did it happen? (2) When did it happen? And (3) how did it happen?

Where?

As far as the kinds of chemical reactions we want to explore are concerned, the first cell had to arise in a liquid water environment. When people first started to imagine scenarios for the origin of life, one of the more popular theories went by the name of the "primordial soup" hypothesis. It worked this way: after the oceans had formed, some type of Urey-Miller reaction took place in the atmosphere and organic molecules rained down into the ocean, where they interacted with one another. This chemically enriched ocean was the primordial soup—in fact, the Smithsonian Institution made a film of TV chef Julia Child mixing, with her characteristic flair, the ingredients of a primordial soup in a pot. A variety of theories were put forth to explain how the first cell formed out of that primordial soup—theories that include the following:

- *Random combinations in the open ocean.* According to these theories, the extreme unlikelihood that a chance encounter between molecules would produce something like a cell is overcome by the fact that the primordial soup could have been around for hundreds of millions of years, allowing lots of time for unlikely combinations to arise. Once a reproducing cell appeared, natural selection would take over, and available energy and material resources would quickly become unavailable for a second origin of life. This sort of explanation, in which a random event comes to dominate life, is known as a "frozen accident" theory.
- *Random combinations in fat globules.* This is similar to the above theory except for its contention that molecules were interacting inside fat globules formed by a Urey-Miller process. This solves the problem of how to make a cell membrane. Each globule can be thought of as a separate experiment in the origin of life.
- *Tidal pools.* When the primordial soup sloshed into a tidal pool, the water evaporated, leaving heavier organic molecules behind. Thus, tidal pools were places where organic material in the soup could be concentrated, a process that speeds up chemical reactions. Tidal pools are as close to Darwin's warm little pond (see the preceding chapter) as we can get, and for a long time they were the leading contender for the place where life on Earth originated.

In the last half of the twentieth century, a hypothesis regarding a new watery birthplace for life began to gain popularity, as we've already mentioned. To the surprise of many life scientists, thriving ecosystems were found around sea vents in the deep ocean. These are places where hot magma comes to the seafloor surface, creating regions of hot water at high pressures and bringing up a rich stew of chemicals that serve as an energy source for a complex collection of life. Given that these deep-sea vents can support life now, the reasoning goes, why shouldn't they have been the place where life first appeared? After all, the deep sea is shielded from the Sun's ultraviolet radiation— radiation that could easily destroy newly built molecules at the surface. The idea is that life first emerged in this sheltered environment and only later migrated to the surface. The authors guess that a slim majority of origin-of-life scientists would pick deep-sea vents as the place where life originated if they were asked for their best conjecture today.

When?

The solar system formed about 4.5 billion years ago, and for the first 500 million years or so the process of planetary formation produced a steady rain of massive meteorites on the nascent Earth. Even after the collision with the Mars-sized object that led to the creation of the Moon, the Late Heavy Bombardment continued. If an Ohio-sized chunk of rock fell from the sky, it would have vaporized any oceans that had formed, sterilizing the planet for thousands of years, and our current theories suggest that there would have been many such collisions in our planet's youth. Thus, any life that arose before the bombardment ended would have disappeared without a trace.

We think that the last such big impact occurred about 4 billion years ago, and we can mark this as the opening of a window in which life could have started to develop. On the other hand, we have fossils of fairly complex life (think "pond scum") dated to about 3.5 billion years ago, and the consensus is that life started soon after the bombardment stopped. There is also indirect evidence, based on the abundance of carbon isotopes, that suggests that life may have started as early as 4.1 billion years ago. Thus, we can say that life on Earth appeared as soon as the geological and astronomical conditions allowed for it. And this, of course, is a conclusion that is important for the development of life on exoplanets.

How?

Once we come to a decision about where and when life on Earth originated, we are immediately confronted with the task of explaining how the complex chemistry of life came into existence. This presents us with a kind of chicken-and-egg dilemma. Modern living systems run a variety of chemical reactions, usually with large, complex molecules. In order for these reactions to proceed at a rapid rate, other complex molecules, called enzymes, must be around to make the reactions go. An enzyme does not take part in the reaction itself, but the reaction cannot take place without it. It is something like a real estate broker, who doesn't buy or sell a house but brings the buyer and seller together to make the sale happen. In modern living systems, the enzymes are proteins whose structure is coded in DNA. And this is where the problem arises—you need the enzymes to decode the DNA, but you need the DNA to make those enzymes before the process starts.

A variety of schemes have been dreamed up to get around this problem. In some theories, the code for the initial enzymes came from a nonliving source—static electricity on clay surfaces, for example. Most scientists, however, are investigating one of two major modes of explanation, which we can characterize as RNA World and Metabolism First.

RNA is a molecule that in modern systems is involved in turning the code carried in DNA into the protein enzymes that run chemical reactions in living systems—in effect, it carries the instructions that tell various enzymes how to string the amino acid "beads" together to make a protein. In the early 1980s, however, scientists discovered that RNA could also act as an enzyme—chemists Thomas Cech and Sidney Altman received the 1989 Nobel Prize for this work, in fact. This dual nature of the molecule opens an intriguing possibility, because it means that RNA can both carry the instructions for making a protein and act as the enzyme needed to carry out the actual assembly. In terms of our analogy, RNA can be both the chicken and the egg.

RNA World scenarios, then, go something like this: in a primordial soup–type environment, one created by Urey-Miller reactions or by material brought in by comets or asteroids, random molecular interactions produce RNA molecules that in turn produce enzymes that run the chemical reactions that allow the cell to grow and multiply. The RNA, in its role as enzyme, produces the molecules needed by the newly formed cell, while in its role as a code carrier, it ensures that the right proteins are made to carry out these reactions. Eventually, the full panoply of the modern cell, with DNA carrying the code and RNA converting that code into protein enzymes, develops through the process of natural selection.

The key point about RNA World, then, is that it can be classified as a frozen accident theory, a theory that depends on the chance assembly of a rather complex RNA molecule as a first step.

Metabolism First theories start from a different point of view. In essence, they reject the notion that since life is complex now, it must have been complex at the beginning—the notion that there has to be a yawning gulf between living and nonliving systems. A standard analogy is to compare living systems to the interstate highway system. Today that system is quite complex, with thousands of miles of paved highway, gas stations, support industries for gasoline, tires, automobiles, and so on. We know, however, that the system didn't start complex—it began with pre-Columbian game trails and progressed to wagon roads and simple paved highways before developing into the interstate system. In the same way, Metabolism First theorists argue, life began with simple chemical reactions—reactions that can proceed without enzymes—and developed into its current complexity over time.

Physicist Eric Smith, then at the Santa Fe Institute, put the Metabolism First approach into context by posing a simple question. "What was the problem in the early Earth," he asked, "that was solved by the development of living systems?"

Another analogy: Imagine that there is a pond of water on top of a hill. We know that the system "wants" (if you will excuse the anthropomorphism) to move that water to a lower-energy state at the bottom of the hill. This is the problem. The solution is for the water to cut a channel in the hill and flow down.

In the same way, to take one example, geochemical processes in the early Earth produced hydrogen and carbon dioxide. In this situation, the hydrogen wants to give up an electron and the carbon dioxide wants to accept it, but this reaction proceeds slowly. In effect, the electrons are stuck on top of a geochemical hill. The development of life provides a channel to get the electron down the hill. Thus, the appearance of life becomes an unsurprising result of the basic laws of physics and chemistry.

Current theories identify what is called the reverse Krebs cycle—also known as the citric acid cycle—as the actual chemical mechanism by which this happens. The cycle involves only 11 relatively simple molecules and does not require enzymes. The cycle takes in energy and uses it to build larger molecules. In modern living systems, the cycle operates in the opposite direction, breaking down larger molecules and producing energy for the cell. It operates at the core of the metabolic process for every living thing on Earth.

As with RNA World, the Metabolism First idea hypothesizes that simple cells (we could even call them protocells) were produced in the beginning, and the full complexity of modern life developed later, as it did with the interstate highway system. To paraphrase William Shakespeare in *Twelfth Night*, life was not born complex, but had complexity thrust upon it.

Implications for Exoplanets

The important point about the appearance of life on Earth is that it happened very quickly. If Metabolism First theories are right, then life will develop whenever the geochemical situation is appropriate. The galaxy should be teeming with life. Frozen accident theories such as RNA World, however, require a separate unlikely chance event—in effect, a separate miracle—for each exoplanet. Given the long time frame and the huge number of exoplanets involved, life might still be plentiful in the galaxy, but that statement is less certain. In any case, RNA World and Metabolism First give us two possible modes by which life "like us" could have developed elsewhere. Nor is it impossible that we will find examples of both on different worlds—the answer to "RNA World or Metabolism First?" may well be "yes." Given our past experience with exoplanets, however, we should also expect that these modes will be only a couple of paths to life among many others that surely exist out in the galaxy.

12

THE SEARCH FOR
EXTRATERRESTRIAL LIFE

Seek, and ye shall find.

Luke 11:9

Discussions of the search for extraterrestrial life usually proceed under the (often unspoken) assumption that we are looking for life "like us," so the discussion ultimately comes down to the question of where we can find liquid water. The reason is simple: life as we know it is based on the chemistry of complex molecules containing large amounts of carbon, and these types of reactions take place most easily in liquid water. All life forms on Earth require liquid water, at least at some point in their life cycle. These facts explain the attention to the Goldilocks planet and continuously habitable zones that tends to dominate the discussion of extraterrestrial life.

As far as the importance of carbon is concerned, some simple chemistry can be applied to the argument. Carbon is an atom capable of forming multiple and strong chemical bonds and thus of making large, complex molecules, especially of the type that has carbon atoms lining up as strongly bonded chains—it's much more able to do this than any other element in the periodic table. Consequently, most scientists (the authors included) subscribe to some extent to what we have called carbon chauvinism. By this, we mean that life like us is sustained by running chemical reactions involving complex carbon-bearing molecules, although not necessarily the same set of molecules that is involved in terrestrial life.

As we outlined in chapter 2, the search for extraterrestrial life in our own solar system has had a checkered history, with hopes declining through the first half of the twentieth century and reviving somewhat afterward. Until the past decade, hopes for a rapid discovery of extraterrestrial life centered on Mars. However, in recent years the subsurface oceans on Europa (the innermost Galilean moon of Jupiter) and Enceladus (a small moon of Saturn) have become the most intriguing places in the solar system to look for extraterrestrial life. Europa is known to have a deep and salty liquid water ocean beneath an ice crust. Salty water mixed with simple and complex organic compounds has been detected spewing out from the interior of Enceladus. In fact, the ejected compounds show evidence of being produced by hydrothermal vents. So over the past decade, the focus of the search for extraterrestrial life in the solar system has moved outward away from the Sun.

The Search for Life on Mars

Mars is the most Earth-like of all the planets in our solar system, and so, as we pointed out in chapter 2, it has received the lion's share of attention as a possible home for life. Over the years a veritable flotilla of orbiter and lander spacecraft has been launched toward the Red Planet, a flotilla that has contributed a series of ups and downs to the search for life.

The first great surprise was provided by the flyby of the *Mariner* 9 spacecraft in 1971. It sent back the first close-up views of the Martian surface, showing river valleys, floodplains, and other evidence suggesting that liquid water had flowed on the surface in the past and opening the possibility that such flows might still be going on, if only sporadically, today. The structures looked much like the river valleys you can see from an airplane window while flying over the western United States. These images had a large psychological impact because they raised hopes that we might find living bacteria when we actually landed on the Martian surface. As an aside, we note that the Valles Marineris, the 4,023-kilometer- (2,500-mile-) long canyon/riverbed system on Mars (the longest valley in the solar system), was named after the spacecraft that first saw it.

Then, in 1976, two Viking spacecraft arrived at Mars, each comprising an orbiter and a lander. The landers, in turn, contained experiments (four on each of them) designed to look for signs of life in Martian soil. Robotic arms scooped samples of that soil into chambers on the landers. One experiment looked for organic molecules directly by measuring the mass of molecules in the sample. This experiment gave a negative result—no organic molecules, no life. Two other experiments that looked for biological activity using radioactive carbon-14 atoms yielded the same result.

Another experiment, however, gave a result whose interpretation is still a subject of debate. The so-called labeled release experiment worked this way: a drop of nutrient mixture was added to the Martian soil. The nutrients consisted of molecules that would be produced by Urey-Miller processes, with some of the carbon atoms replaced by radioactive carbon-14. The idea was that if there were living organisms in the soil, they would metabolize the nutrients and some of the carbon-14 would be detected in the gas above the soil by its radioactive signature.

And that's exactly what happened—the carbon-14 did indeed appear. The problem is that the same sort of results would be seen if there were certain types of chemicals in the Martian soil—chemicals whose presence was confirmed by the Phoenix lander in 2008. Thus the question of whether there are living microbes on the Martian surface does not have a definitive answer at this time. For the record, the authors' interpretation of the Viking results is in line with the general consensus of the scientific community—the presence of life can't be ruled out, but it probably isn't there, at least in the top few centimeters of Martian soil. We now know that hydrogen peroxide saturates the upper levels of the soil, effectively sterilizing it of any life "contamination."

Mars continues to tantalize, however. In 2014, NASA announced that the *Curiosity* rover had detected a transient spike in the abundance of methane (natural gas) as it was ambling across the Martian surface. It wasn't a big signal—the concentration of methane was about 10 ppb (parts per billion), as opposed to about 1,800 ppb on Earth. Methane can be produced by living organisms—indeed, it is commonly captured at landfills and used as fuel on Earth. There are, however, other nonbiological processes that can produce the gas. Once again, we can't rule out a scenario in which subsurface microbes belch out plumes of methane, but we can't rule out other sources, either. Future isotopic measurements of the Mars methane may be necessary in order to determine if it is biological or geochemical in origin.

As far as water is concerned, there is an abundant supply on, and inside, Mars, mostly in the form of water ice—not surprising on a planet with an average temperature of $-55°C$ ($-67°F$). Depending on what scientists assume about subsurface ice, there is enough water that, if it were in liquid form, it would cover the planet with an ocean anywhere from tens of meters to a few kilometers deep. There is even evidence for sporadic flows of liquid at the surface—pictures taken by the Mars Reconnaissance orbiter in 2001, for example, showed gullies in one region where they had not been seen in 1999. Minerals that require liquid water for their formation have been found where these gullies drain out of the sides of impact craters, but direct detection of liquid water has not yet occurred, although in 2015 evidence for a recent flow of what could have been a briny sludge was seen.

The search for life (present or past) on Mars continues. There are plans to launch ExoMars, a joint venture of the European Space Agency and the Russian Federal Space Agency, sometime between 2016 and 2018. This rover will have the capability of drilling down about six feet into the Martian surface, down to the level where scientists think there might actually be liquid water and hence living microbes. NASA is considering a Mars

Rover 2020 project to make a concerted search for biomarkers, perhaps sometime early in the next decade, and, of course, the sample return mission discussed in chapter 2 remains a possible future endeavor.

We've gone through this detailed description of the search for life on Mars to illustrate an important point: finding convincing evidence of the presence of extraterrestrial life might not be easy. Even in a case where we can actually visit a planet, it has proved difficult to nail things down. There are just too many ways to explain the current results without recourse to life. How much more difficult it will be when we go out into the galaxy is a question for us to consider carefully.

Out into the Galaxy

The first step in looking for life in the galaxy is to assemble as complete a catalogue of Earth-type planets as possible. This is the task that will be carried out by NASA's Transiting Exoplanet Survey Satellite (TESS), due to be launched in 2017. During its two-year mission, it will monitor 500,000 bright stars in our vicinity, and it is expected to find several thousand Earth-sized and super Earth–sized planets. These newly discovered planets will be analyzed by other telescopes, such as the James Webb Space Telescope, due to be launched in 2018. Unlike Kepler, which probed only a pencil-thin sliver of the galaxy, TESS will do a full-sky survey, so we will really have a complete catalogue of nearby Earth-type exoplanets.

Once we have that catalogue, the next question is how we will go about searching those exoplanets for life. There is no chance we'll be able to send the kind of probes to them that we have been sending to Mars, so we'll have to fall back on a technique that astronomers have been using since the nineteenth century: spectroscopy.

The basic idea behind this technique is that every atom or molecule that emits or absorbs light or some other form of electromagnetic radiation does so in a specific pattern. If you think of an atom as consisting of a nucleus with electrons in orbit, then when an electron moves from a higher orbit to a lower one, it emits a bundle of light whose energy (and hence color) corresponds to the energy difference between the orbits. Similarly, the atom absorbs light of a given color when the electron moves from a lower to a higher orbit. Since atoms of a specific element have a unique arrangement of orbits, the pattern of light emitted is unique for each element or molecule—you can think of this pattern as a kind of fingerprint allowing you to identify the atom or molecule that produced it. It is customary to refer to bright colors (for emission) and dark regions (for absorption) as "spectral lines" and to the entire collection of lines as a "spectrum."

Once this light is emitted, it makes no difference how far away the observer measuring it is. He or she can be a few feet or a few light-years away—the light pattern is the same. Thus, the technique of spectroscopy, which has been developed to a high level of sophistication and precision, can be used to probe the chemical nature of the atmosphere of exoplanets. This is the only direct information we are likely to get about the chemical composition of an exoplanet.

The main technical problem in applying spectroscopic analysis to exoplanets is separating the spectrum created by the planet from that created by its star. As we mentioned earlier, a simple way to do this is to wait until the planet moves behind the star and observe how the spectral lines change when the planetary atmosphere is no longer making a contribution. In some cases, it is also possible to probe the exoplanet directly by observing it after the transit, so that the telescope can isolate the planet when it is farthest from the star. In any case, the end result will be a compilation of all the atoms and molecules in the planet's atmosphere.

Then what? What should we look for to indicate the presence of life? If we're talking about life like us, we can answer this question. We will need to see evidence of water, for starters. Furthermore, we know that living systems like us require certain key elements—carbon, hydrogen, nitrogen, oxygen, phosphorus, and sulfur (a useful mnemonic is CHNOPS). The presence of these elements—they're pretty common in the galaxy—would signal the possibility that life exists.

If we take the history of life on Earth as a guide, what we would look for as evidence of life would change over time. The earliest life forms existed in an environment that had no oxygen, and it took over a billion years for them to add enough oxygen to the atmosphere for it to be detectable. The spectroscopic signature of a planet that had life in this early stage, then, would be different from that of a planet where life was fully developed.

The first step in our search for life, however, would undoubtedly be to check for the presence of oxygen. It is believed that some of the earliest fossils on Earth were from cyanobacteria that gave off oxygen as a waste product of photosynthesis, so we can assume that the presence of oxygen is indicative of a "green pond scum" planet. On Earth, life was limited to single-celled organisms (the pond scum) for billions of years. Our guess is that most planets where we find evidence for life will be of this type. At the moment, scientists cannot think of a way that significant oxygen can get into an atmosphere without the presence of life, so this could be a definitive test. On the other hand, it's not hard to imagine scenarios in which, when oxygen is detected on an exoplanet, someone suggests a possible nonbiological source for it.

Another, and perhaps better, approach would be to look for combinations, or sets, of molecules that are rare except when they are associated with life. The existence of methane on Mars is significant because methane is continually destroyed by chemical reactions in the atmosphere. Methane cannot exist in the Martian atmosphere unless there is a continual source of renewal, which would presumably be underground, to balance that chemical destruction. So the combination of carbon dioxide and methane is what is called a disequilibrium mixture. On Earth, that same disequilibrium is produced by life through the release of methane as a by-product of metabolism. Methane, in this case, is a biomarker of life. It is actually possible to determine if the methane on Mars is from biology or geochemistry by looking at the ratio of the heavy to light isotopes of carbon in the methane, but the observational determination of that isotopic signature is exceedingly difficult for Mars, and we are not there yet.

When we look at the chemical makeup of the atmospheres of exoplanets, such disequilibrium mixtures would require something like life to be produced. Two other disequilibrium combinations could be ammonia and oxygen, or molecular hydrogen and oxygen. Although the existence of disequilibrium mixtures is not proof of life, it would tell us that it is worth looking deeper into the exoplanets on which they are found.

In any case, this is about as far as we can go in the search for life unless we find we are dealing with an advanced civilization—a subject we discuss in the next chapter. It is not hard to imagine an exotic life form that could put other kinds of atoms and molecules into its planet's atmosphere, but, as our experience with Mars shows, it's also possible to imagine alternative processes that would produce the same result without the presence of life. Our sense is that actual proof of simple life on the surface of an exoplanet may ultimately be extremely difficult to obtain. But, as physicist Richard Feynman (1918–88) once said, "Much more can become known than can be proven."

What about Exomoons?

Up to this point, we have confined our discussion to exoplanets, but the experience with Europa in our own system should convince us that there is another important category to consider in the search for life, and that is the moons that might circle those planets. There are, in fact, two scenarios in which an exomoon might produce the conditions for life even when its parent planet does not:

1. The moon belongs to a Jupiter-type planet that is in the CHZ of its star. In this case, our instruments might tell us that the planet is unsuitable for life, but in the right situation, the moon could have liquid water on its surface.
2. The moon belongs to a Jupiter-type planet far from the CHZ but, like Europa, is heated by tidal flexing or another

heat source and has a subsurface ocean. It's hard to see how we could get evidence of such an ocean from a distance, but we have to keep the possibility in mind.

For the record, there is already some tentative evidence of exomoons in the Kepler data set.

Are We Sure We Would Know It If We Saw It?

It cannot have escaped the reader's attention that in our discussion of life in the previous chapter and this one, we have talked almost exclusively about life like us—that is, life based on the chemistry of carbon-based molecules. The reason for this is simple: since we don't know about any other kind of life, we have no idea what we should look for as evidence of life that is *not* like us. Think of this as another instance of the curse of the single example.

It has become commonplace in the astrobiological community to use the quote from Justice Potter Stewart—the one about knowing pornography when he sees it (see chapter 4)—to deal with the problem of recognizing life that is not like us. Before we move on to discussing extraterrestrial intelligence (as opposed to extraterrestrial life), we would like to take a moment to raise the question of whether we would really know life if we saw it. Let us consider just two examples.

In chapter 2, we talked about Titan, the principal moon of Saturn. Titan has liquid oceans—not liquid water, but liquid methane. At the frigid temperatures on Titan, any chemical reactions, whether they involve carbon or not, must proceed very slowly. So ask yourself this: if there were a life form on Titan that took 1,000 years to draw one breath, would we recognize it as alive or think it was just a rock? Would we know it when we saw it?

Or imagine a planet such as one invented by the science fiction writer Isaac Asimov (1920–92), in which each little piece— each rock, each grain of sand—was connected to every other piece. Each piece by itself would be unremarkable, but taken together they would be a highly advanced life form. Looking at a single rock would be something like looking at a single transistor and trying to deduce the function of the supercomputer of which it is part. We doubt that we would recognize this type of life form if we saw it, either.

We could go on—science fiction writers have imagined all sorts of exotic life forms that might exist, and, frankly, there is no way we could devise a testing protocol to screen for every imaginable kind of life. We will undoubtedly begin our search for extraterrestrial life by looking for something like us. Better, after all, to begin with the familiar. But we should always have in the backs of our minds the possibility that the galaxy might throw something totally unique and unexpected at us—something we can't even imagine right now. Our experience, based on the extremely limited robotic exploration of our solar system, is that we are continually surprised at how complex even "simple" planets and moons in our own backyard can be. We should be prepared for even larger surprises as we explore our more distant neighborhood.

13

THE SEARCH FOR EXTRATERRESTRIAL INTELLIGENCE

You miss 100 percent of the shots you don't take.

Hockey great Wayne Gretzky

L et's face it—while the discovery of a green pond scum planet or even the remains of fossil life on Mars would be a major breakthrough for the life sciences, it isn't going to generate a lot of popular enthusiasm. We're afraid that decades of science fiction have conditioned people to expect intelligent, technologically sophisticated extraterrestrials, not just cyanobacteria. And that's why public attention remains focused on the search for extraterrestrial intelligence (SETI).

The scientific attitude toward SETI has an interesting history. It's clearly been on scientists' minds—in the early twentieth century, for example, Guglielmo Marconi (1874–1937), the inventor of radio, reported receiving radio signals "from Mars," and we're sure that his contemporaries enjoyed the era's science fiction—but there was a general sense that an undertaking such as SETI was outside the realm of serious science. Researchers just didn't have the tools to do it.

All of that changed in 1959. In that year, the prestigious British journal *Nature* published an article by physicists Philip Morrison (1915–2005) and Giuseppe Cocconi (1914–2008) arguing, in effect, that if there were an extraterrestrial civilization out there that was trying to communicate with us, our newly constructed radio telescopes would enable us to listen to what they had to say. In one fell swoop, this paper moved SETI from the domain of science fiction to the realm of testable science.

The Green Bank Conference and the Drake Equation

Following the *Nature* paper, in 1961, a small conference was held in Green Bank, West Virginia, which was the home of the world's most advanced radio telescope at the time. Eleven people from many branches of learning got together to try to estimate the number of intelligent life forms in the galaxy that might be attempting to communicate with us. Despite its small size, the conference had an enormous impact in shaping the discussion of SETI for the next half century.

The main result of the Green Bank conference was an equation that specified the kind of knowledge we'd have to acquire to move the SETI discussion forward. Called the Drake equation, after astronomer Frank Drake, it has this form:

$$N = R \cdot f_p \cdot n_e \cdot f_l \cdot f_i \cdot f_c \cdot L$$

Its symbols have the following meanings:

- N = number of extraterrestrial civilizations trying to communicate with us right now
- R = rate of star formation
- f_p = probability that a star has planets
- n_e = number of Earth-type planets in a planetary system
- f_l = probability that life will develop
- f_i = probability that life will develop intelligence, given that life has developed
- f_c = probability that there will be a technological civilization capable of sending signals
- L = length of time signals will be sent

To emphasize the importance of this equation, we begin by analyzing its terms as they would have been seen in 1961; then we turn to the question of how the Drake equation is analyzed in the light of modern knowledge. Finally, we examine how it should be seen in the light of our new understanding of exoplanets.

At the time of the Green Bank conference, people believed that the galaxy had about 10 billion stars and was about 10 billion years old. Hence, R was taken to be 1 (i.e., one new star, on average, formed each year). Values up to 10 were considered acceptable on the grounds that star formation was probably more common in earlier eras.

Given the nebular hypothesis, planetary systems were believed to be fairly common, although in 1961 there was no direct evidence of this. Consequently, f_p' was assigned a value between 0.5 (half the stars have planets) and 1 (all stars have planets).

With the next term, the number of Earth-type planets, we begin to see how recent knowledge can play an important role in SETI. Looking at our own solar system, the Green Bank participants knew that this number had to be at least 1 (Earth), and at the time it wasn't considered unreasonable to think life existed on Mars and maybe even Venus. Toss in a few of the Jovian moons and any number between 1 and 5 could be defended for n_e.

From this point on, the Green Bank participants relied heavily on the Copernican principle (also known as the principle of mediocrity). It holds that there is nothing special about Earth—that what happens here is typical of what happens elsewhere in the universe. Thus, f_l and f_i, the probability that life and intelligence developed, were taken to be 1. This was, remember, before anyone had the idea that there might be such a thing as a continuously habitable zone—an idea that, as we shall see, greatly reduces the acceptable values of both n_e and f_l. It was also before the time that people took the notion of a green pond scum planet seriously. In 1961, after all, the fossil record of life on Earth seemed to start about 500 million years ago, in what was called the Cambrian explosion. The discovery of the record of single-celled life going back 3.5 billion years was still in the future.

The value to use for f_c, the probability of a civilization capable of interstellar communication, was a subject of debate. What the conferees did was to count civilizations in human history—Egyptian, Greek, Chinese, Roman, Aztec, etcetera—and note that only one—ours—had built radio telescopes. This gives a value of f_c between 1/5 and 1/10, depending on how you count.

The reader will undoubtedly have noticed that, as we proceed from left to right in the Drake equation, our ability to assign numbers to the entries is characterized by greater and greater uncertainty. Nowhere is this more evident than when we consider the last term, L, the length of time an extraterrestrial civilization will continue signaling. This necessarily involves speculations in a field we could call "exosociology"—a field that the authors fervently hope doesn't really exist.

In any case, the Green Bank participants looked at two extreme cases. On the one hand, they knew that human beings had been sending radio signals out into space since the nineteenth century. These signals weren't sent deliberately, of course, but radio and TV broadcasts travel in all directions, out into space as well as to their intended targets. It is, in fact, a matter of dread for some intellectuals to think that the human race may be announcing its presence to the galaxy in an outgoing wave of *I Love Lucy* reruns. The participants assumed 100 years to be the lower bound for L.

The upper bound is harder to pin down. What the conference participants did was to substitute the question "How long *could* a society communicate?" for the question "How long *would* a society communicate?" At one level, we might argue that it would be possible for a civilization to send signals for as long as its star was shining. For Earth, this would be billions of years—a ridiculous number. Instead, the Green Bank participants picked an estimate of 100 million years—a typical geological time frame. This is, for example, a typical time frame for plate tectonic movements on Earth, a typical time frame for a mountain chain to be eroded away, and so forth. It is also roughly the time between large meteorite impacts on Earth—the kind that may have wiped out the dinosaurs. This still seems to be a very large number to the authors, but that's what was used at the conference. Thus, L was taken to be between 100 and 200 million years.

Depending on how optimistic they wanted to be in choosing numbers for the Drake equation, the conference participants were able to report large numbers for N—numbers in the millions or even higher. The media seized on this result. People talked about humanity joining the "Galactic Club," which was imagined to be a huge group of advanced civilizations out there waiting to welcome us.

Things didn't turn out that way, of course—an outcome we'll look at in some detail in the next chapter. Before that, however, let's look at how we might work out the Drake equation in the light of modern knowledge.

The Drake Equation Today

There are a few minor points we can change in the above analysis. For example, as we pointed out in chapter 9, we are discovering that there are many more small M-type stars in the galaxy than people knew about in 1961. This means that values of R in the range of 20–40 are more reasonable than values in the range of 1–10—not a big difference.

On the other hand, the development of the notion of a continuously habitable zone (CHZ)—a zone around a star in which liquid water can stay on the surface of a planet—has a huge effect on the value we assign to n_e, the number of Earth-type planets around a star. There are several points to make here. First, if we think that surface oceans are necessary for the development of life, then the planet we're looking at has to be in the CHZ of its star. Furthermore, modern calculations of atmospheric development suggest that oceans will only survive on planets not too different in mass from Earth. Thus, the value of n_e should not be concerned with Earth-type planets, but with "Earth-mass planets in the CHZ of their stars." Based on the data available from Kepler, this number is going to be much smaller than the 1–5 used at the Green Bank conference—perhaps as small as 1/100 or even less.

In addition, our discovery of subsurface oceans on the moons of the Jovian planets opens a new line of inquiry—think of it as a separate branching in the Drake equation. In effect, the term n_e in the equation will have to be replaced by a term such as $n_{CHZ} + n_{SS}$, where the first term represents Earth-sized planets in the CHZ of their stars (as well as moons of any planet in the CHZ) and the second represents the number of moons with subsurface oceans. Aside from our own solar system, we have no data whatsoever on the value of n_{SS} in the galaxy as a whole. Also, the notion that life could exist on planets roaming between the stars would require yet another addition to this term. However, the observational study of such rogue planets is too young for us to make a credible assignment of a value for such a term.

Although we know a good deal more about the origin of life on Earth than we did in 1961, as outlined in the last chapter, the new knowledge really doesn't help us much in assigning a value for f_l, the probability of life developing. The one thing we do know is that life developed very quickly on Earth once the conditions were right. Some scientists have even suggested that life developed many times during the Late Heavy Bombardment, only to be wiped out and have to start over again after each impact. Furthermore, the kind of prebiotic chemistry we discussed in the previous chapter, particularly the chemistry around deep-ocean vents, could take place in subsurface oceans on moons as well as on planetary surfaces. We therefore, faute de mieux, assign f_l a value of 1, as did the Green Bank conferees.

Intelligence

This is a hard word to define precisely. Take life on our own planet as an example. We would clearly want to say that humans are intelligent, and we would probably apply that adjective to other types of vertebrates—mammals, birds, reptiles, amphibians. But what about fish? Lobsters? Clams? Pumpkins? Drawing a line can (and does) lead to heated debates among scholars, but fortunately the exact location of that line doesn't matter much for our current analysis. The key point is that the one statement everyone would agree on is that intelligence requires multicellular life. And this, in turn, forces us to pay attention to an important evolutionary event—the development of eukaryotes.

A bit of explanation: *karys* is the Greek word for "nucleus" or "kernel." There are two types of cells on Earth: those whose DNA is contained within a nucleus, and those in which the DNA floats freely. The former are called eukaryotes ("true nucleus") and the latter are called prokaryotes ("before the nucleus"). All the cells in your body—and all the cells in every multicellular organism on the planet—are eukaryotes. Thus, if we want to talk about the evolution of intelligence, we have to understand something about how these types of cells came into existence.

The first cells that developed on Earth—presumably similar to the cyanobacteria that produced the green pond scum planet—were simple prokaryotes, as were the cells that followed them for over a billion years. About 2 billion years ago, however, an extraordinary string of events started to occur. Two prokaryotic cells discovered that there was an evolutionary advantage to being incorporated into a single cell—an advantage that wouldn't exist if they remained separate. As a result, the two simple cells came together to form a more complex cell. This process, endosymbiosis, eventually led to the complex system that is the modern eukaryotic cell, in which many substructures (called organelles) perform many of the cell's functions. To take one example, all eukaryotic cells have organelles called mitochondria whose function is to generate the cell's energy. Modern mitochondria have two cell membranes, a fact that scientists interpret by saying that one membrane came from the original prokaryote that was absorbed and the other came from the membrane of the original host cell. What this means is that to assign a value to f_i, we need to have some idea about the likelihood of endosymbiosis occurring.

We should point out that the advent of eukaryotes did not lead to the demise of prokaryotes—evolution just doesn't work that way. Eukaryotes were able to move into new ecological niches (by utilizing oxygen, for example), leaving the prokaryotes to stay in their old ways.

There are two ways of looking at endosymbiosis. On the one hand, it did occur, probably repeatedly, on Earth. On the other hand, it took a long time to happen—over a billion years. The question then arises as to how easily it will happen on exoplanets. If a billion years is a typical time scale, then there will have been plenty of time for it to occur elsewhere. If, however, we were lucky and endosymbiosis happened earlier here than normal, it could be that there are many green pond scum planets out there waiting for eukaryotes to develop.

Once there are eukaryotes, there is another step that has to take place before intelligence can develop, and that is the step in which cells discover an evolutionary advantage in being part of a multicellular organism. As mentioned above, we used to think that life appeared about 500 million years ago, in the Cambrian explosion. We now understand that what happened in the Cambrian was the development of organisms with hard parts that are easy to fossilize. In fact, we now know that multicellular organisms without hard parts (think jellyfish) were present at least 800 million years ago.

So once again, we have a puzzle. The development of multicellularity (and hence the possibility of intelligence) took over a billion years to happen, which can either mean that it will happen elsewhere given enough time or that it is rare. In the latter case, we might have planets with lots of single-celled eukaryotes but no intelligent life. Once again, for lack of any reason to do otherwise, we'll assume that Earth is typical and assign f_i a value of 1, with the understanding that it could be much smaller. We will also assume, since we can see no reason to do otherwise, that there is no difference between surface and subsurface oceans in this process.

All of which brings us to the last two terms in the Drake equation. The argument that the development of intelligence must lead to a technological society doesn't seem very strong to us. After all, dinosaurs—arguably intelligent animals—ruled Earth for over 200 million years without ever developing a technology. There could easily be a lot of "dinosaur planets" out there.

In addition, even if intelligent life and a technological civilization developed in a subsurface ocean under a layer of ice many miles thick, it seems to us extremely unlikely that a science such as astronomy—much less the ability to build a radio telescope—would develop quickly. For the sake of argument here, we drop considerations of subsurface oceans as far as SETI is concerned from this point on. As far as a numerical value for f_c is concerned, we can't think of a better way to proceed than the culture-counting method used at Green Bank, so we'll take its value to be 1/5 to 1/10.

The final term in the equation is, of course, the most uncertain. Going on human experience, we feel that 100 years is a reasonable lower bound for L, the length of time signals will be sent. But human experience also suggests that the upper bound will be set by social and institutional forces, rather than geological or astronomical ones. Few human institutions have lasted more than 1,000 years—even major religious systems seem to die out in this kind of time frame. We feel, therefore, that an upper bound on L of 1,000 years, or at most 10,000 years, is much more reasonable than the kind of bounds used by the Green Bank conferees.

Putting these sorts of updated numbers into the Drake equation can produce estimates of N that range from a few hundred (a somewhat diminished Galactic Club) down to numbers that are less than 1 (not every galaxy has an intelligent technological life form). The key point is that we could very well be the only advanced technological society in our own galaxy right now. This is an important fact to keep in mind as we move on to discuss SETI programs.

SETI

Starting in 1960, there have been many SETI-type searches, none of which have provided any credible evidence for extraterrestrials. The federal government has funded a few of these searches, but for the most part they have been funded by private organizations. The most recent private effort was a donation of $100 million to a SETI organization called Breakthrough Listen by Russian tycoon Yuri Milner. A typical scheme has been to either buy time on existing telescopes or equip an obsolescent telescope with modern electronic equipment to monitor many stars simultaneously. In all of these searches, however, we can identify two crucial questions: (1) where do you look? and (2) what do you look for?

The first SETI searches tended to look for signals from nearby Sun-like stars. Later, these searches were expanded to many regions of the sky. In the end, the search results have all been the same, regardless of where we look.

As far as specifying what to look for, the original 1959 Cocconi and Morrison paper suggested that the most likely way that an extraterrestrial civilization would try to send a message would be to use the (microwave) frequency associated with a particular change in interstellar hydrogen—a frequency corresponding to a wavelength of 21 centimeters (8 inches). Their reasoning was that this is a ubiquitous wavelength in the galaxy and so represents a "natural" choice dictated by nature herself.

Actually, the problem of choosing what frequency to listen to is something like running through the radio dials in a strange city. The station you're looking for could be anywhere, and only a complete search can be guaranteed to find it. What Cocconi and Morrison were doing was, in effect, choosing a "magic frequency" at which signals would be sent. Having such a frequency simplifies the SETI search immensely, since it avoids the "running through the dial" problem. On the other hand, if the search comes up empty, you have no way of knowing whether it's because you were looking at the wrong frequency or just in the wrong place. Many early SETI searches relied on different choices of the magic frequency—the peak of the cosmic microwave background was a popular choice for a while. But whatever the choice, no magic frequency seemed to work.

Since we can't seem to find a magic frequency or a magic place to look, the only alternative seems to be an all-sky, all-frequency search—a daunting task. Just think of the billions of stars in the galaxy and the huge numbers of frequencies that have to be sampled for each one. In May 1999, SETI scientists at the University of California, Berkeley, introduced a novel tool to deal with this plethora of data. Called SETI@Home, it allows individuals to use their personal computers to analyze SETI data. The idea is that SETI scientists send packets of data to each participant. When the individual computers are not being used for other tasks, they turn to looking for signals in the data (the SETI program is typically used as a screen saver). There are millions of SETI@Home participants in about 100 countries around the world, and this technique of harnessing unused computer capability has been employed in other areas of science as well.

Scientists have thought about other possible signatures of an extraterrestrial civilization besides radio signals. Perhaps the most interesting of these is the so-called Dyson sphere, an idea introduced by physicist Freeman Dyson. He thought about how a truly advanced civilization would obtain energy. His conclusion: the main energy source in any solar system is the central star, and the way to tap that energy most completely would be to construct a giant sphere around that star and to intercept all of its outgoing energy. This is the Dyson sphere—the extraterrestrials would presumably live on the inside surface. An outside observer wouldn't see the star, of course, but the sphere would radiate in the infrared. Thus, the characteristic signature of a Dyson sphere would be an extremely bright, point-like infrared source in the sky.

So, having analyzed the possibility of detecting signals from an extraterrestrial technological civilization or finding indirect evidence of their existence, we come to the single incontestable fact about SETI we mentioned above. Despite a half century of searching, there is not one piece of evidence to suggest that any such civilizations actually exist. Some scholars call this the problem of the "Great Silence." It will be the subject of the next chapter.

Could the Drake Equation Be Completely Irrelevant?

The Drake equation has dominated the SETI debate since the 1960s. Nevertheless, we feel that our new discoveries about exoplanets, together with some new theoretical insights about planetary atmospheres, indicate that it can no longer be used as a reliable guide in the future. It's just too restricted, too bound by the chauvinisms we talked about in chapter 1.

Start with R, the first term in the equation. No matter what value you choose for R, the assumption is that before you can have life, you must have a star. But, as we saw in chapter 7, rogue planets are perfectly capable of supporting life—recall the analogy of houses with the lights turned off but the heating system still running. Even rogue worlds whose surfaces are frozen, such as Pluto (see chapter 8), might have subsurface oceans. Given that there are probably many more rogue planets in the galaxy than there are planets circling stars, at the outset we see that the Drake equation is designed to deal with only a small fraction of the planets out there. We'll return to some implications of the abundance of rogue planets in the epilogue.

Moving to the right in the equation, the way n_e, the number of Earth-type planets, is used today depends on the concept of the continuously habitable zone (CHZ). The CHZ, recall, is the region around a star in which water on a planetary surface can remain in liquid form for billions of years. As we pointed out above, the traditional analysis of the Drake equation ignores the existence of subsurface oceans. It appears, however, that in our solar system—and probably in the Milky Way as a whole—most of the water is not on the surface of planets. This implies that the existence of life is probably much more widespread than is implied by the equation. Little attention has been paid to this fact in SETI discussions.

The importance of the CHZ has also been called into question by recent calculations. Basically, it appears that the CHZ (which itself is a concept based on computer models) can be expanded considerably if we make different assumptions about the composition of the planetary atmosphere. Think of our own concern about the effects of carbon dioxide in Earth's atmosphere as an example of how such effects might arise. At the moment, all we can say is that this added complication makes the CHZ a much less reliable tool in SETI.

Finally, let's go on to the last term in the equation, L, or the length of time a civilization will send signals. While the image of a sphere of *I Love Lucy* reruns expanding at the speed of light is an amusing one, it doesn't reflect the modern state of our technology. Broadcasting waves into space is an extraordinarily wasteful process. On Earth, an increasing amount of our communications are carried by optical fibers or by focused beams sent to orbiting satellites. Neither of these will add to our radio signal as seen from other planets. Consequently, the lower limit of 100 years we set on L could well be an upper limit as well. It could, in other words, represent the time between the discovery of radio waves and the discovery of optical fibers. In this case, radiation leaking from a technologically advanced civilization would be detectable for only a short time. We would only see them, in other words, if they were actually trying to contact us—a difficult outcome to predict.

Should We Continue to Search for Extraterrestrial Intelligence?

Given the failure of all SETI searches to date, the question naturally arises as to whether we should keep looking. We approach the question this way: there are few experiments in science that are guaranteed to give a significant result no matter how they turn out. SETI is one of them. If . . .

SETI searches turn up another technological civilization, fantastic!

SETI searches verify that we are the only advanced life form in the galaxy, fantastic!

And, as Wayne Gretzky reminds us in the quote at the beginning of this chapter, the only way a SETI effort can fail is if we don't do it. So, hell yes—we should do SETI.

14

THE FERMI PARADOX

Where is everybody?

Physicist Enrico Fermi

The story is that it all started one day in 1950, when a group of prominent physicists—all veterans of the Manhattan Project—were walking to lunch at the Fuller Lodge in Los Alamos. They were discussing the spate of recent UFO sightings that had been claimed in the area, and the conversation turned to the topic of extraterrestrial civilizations. Out of the blue, Enrico Fermi (1901–54), a man well known for his ability to see to the heart of a problem, asked a simple question: Where is everybody? In the years since then, scientists have come to realize that Fermi's offhand question is, in fact, the deepest question we can ask about life in our galaxy. The fact that there is no evidence for the existence of extraterrestrials in spite of the calculations suggesting that they should exist is known as the Fermi paradox.

Born in Italy, Enrico Fermi quickly rose to prominence in the new science of nuclear physics. His pioneering work in what is now called the "weak interaction" was recognized by his being awarded the Nobel Prize in 1938. That was not a particularly happy year for Fermi, however, since the fascist government in Italy introduced laws that threatened his wife, who was Jewish. Consequently, he joined the great wave of scientists fleeing Hitler's Europe, winding up at the University of Chicago. There, in 1942, he operated the first nuclear reactor under the stands of the old football stadium (a building that has since been replaced by a library). He went on, as intimated above, to play a prominent role in the Manhattan Project and as an advisor to the postwar Atomic Energy Commission. Since his untimely death, he has been honored by having many important facilities named after him—the Fermi National Accelerator Laboratory, the Fermi Gamma-ray Space Telescope, several reactors, and the artificial element number 100 (fermium). His element is located between einsteinium and mendelevium, which gives some notion of his status among scientists.

So why has his offhand question played such an important role in the debate about extraterrestrials? To understand this, we can go back to our old device of compressing the lifetime of the universe into a single year. In this scheme, the Sun and our solar system formed in the late summer (Labor Day is a convenient approximation), modern humans showed up a few minutes before midnight on New Year's Eve, and all of recorded history took place while the ball is descending in Times Square, with modern science appearing in the last second of that descent.

The point is this: if there really are other technological civilizations out there, it is extremely unlikely that they developed science after we did—after all, they had the whole year to discover the laws of nature. To understand what follows from this statement, let's look at a possible future for the human race.

We'll start at Princeton University in the 1970s, where physicist Gerard O'Neill (1927–92) was teaching a seminar centered around an interesting question: is the surface of a planet really the best place for a technological civilization? The answer the class came up with was "no," and from their deliberations came the design for a structure now called an O'Neill colony.

Imagine a hollow doughnut, a mile or more across, rotating slowly in space. In O'Neill's vision, people live inside the doughnut, and the centrifugal force associated with its rotation substitutes for gravity. Using solar or nuclear power, possibly with ancillary doughnuts for raising crops, such a system could be self-sustaining, a true move of humanity away from our home planet. It is almost within our technological capabilities to build such a structure right now, if not within our budgets. In any case, we should expect that any extraterrestrial race that has come to our level of technical sophistication should also be able to build something like an O'Neill colony.

Let's imagine how something like O'Neill colonies might play out in our future. Eventually, we can expect that people in colonies like this would leave the space around Earth and move to the truly prime real estate in the solar system, the asteroid belt, where ample material and solar power are available.

It's the next step that has enormous implications for the Fermi paradox. After a few generations have spent their lives in something like an O'Neill colony, will it really matter to them if their colony is on the way to another star system rather than in the asteroid belt? As the best locations in our own system fill up, it is reasonable to suppose that future space colonists will follow the example of their forebears and "light out for the territories," except that, in this case, that would mean moving to other solar systems. In essence, we suggest that they would turn their colonies into interstellar starships. How hard would that be?

Let's make two extraordinarily conservative assumptions. Let's assume that (1) there is no way to get around the speed-of-light barrier—no "warp drive"—and (2) no major technological advances will be made in the next couple of centuries. The immense distance between stars would require travel times of a century or more, which would mean that the starship would be multigenerational—you get on, your grandchildren get off. Several propulsion systems for such a trip have been proposed—for example, one in which the ship scoops up rarefied interstellar hydrogen to run its power and propulsion systems. The idea of such a multigenerational starship is already a staple of science fiction. In the epilogue, we will return to the discussion of colonization and see how this scenario might be influenced by the presence of rogue planets.

The point of this exercise in futurology is that once a civilization reaches our level of sophistication, it is only a matter of a few centuries before it can start colonizing other star systems. If we can imagine ourselves doing it, then there's no reason extraterrestrials couldn't do it as well. The important point for our discussion is that we are talking about a time span of only a few hundred years. In terms of our galactic year analogy, this amounts to only one second. Basically, as soon as the ball touches down in Times Square, Earth could be the center of an expanding wave of human colonization. No one would even have time to say, "Happy New Year."

How long would it take that wave to engulf the entire galaxy? Most calculations give times on the order of 30 million years or so. And while this is an extremely long time on a human scale, it is only one day in our galactic year. So if extraterrestrial civilizations have been popping up throughout the galactic year, and if at least some of those civilizations are as scientifically adept as we are, there should have been multiple waves of colonization sweeping over the solar system. If you take an optimistic interpretation of the Drake equation, in fact, over a dozen such waves should be sweeping through the galaxy right now.

So . . . where is everybody?

That, in essence, is a modern look at the question Fermi asked over a half century ago, one we still haven't been able to answer. His point can be stated this way: we shouldn't be looking for extraterrestrials out there, as we do in SETI—we should be looking for them right here. And if we ignore the silliness of UFOs and ancient astronauts, we can say that there is no evidence whatsoever for extraterrestrials being here now or in the past.

Where is everybody? Why the Great Silence?

The Great Filter

William of Ockham, as noted earlier in the book, was an English scholar who is famous for one throwaway line in an otherwise turgid theological treatise. Called Occam's razor, it says, "Plurality must never be posited without necessity." In essence, it tells us that when we have a question to answer, the simplest solution is the one we should choose. The concept shaves away complexity; hence the word *razor*.

There is no doubt that the simplest answer to the questions "Why the Great Silence? Why don't we hear any SETI signals?" is that we don't hear signals because no one is sending them. There are a number of other explanations that have been put forward, and we can look at them briefly before taking William of Ockham seriously. Basically, the explanations can be divided into three categories:

1. They really are out there, but they're not interested in us.
2. They really are out there, but they're protecting us.
3. They really are out there, and we're going to get it unless we mend our ways.

An example of the first category would be a race of extraterrestrials living in a Dyson sphere, happy as clams with their star's energy and supremely uninterested in anyone else. Another possibility, discussed in the epilogue, would be extraterrestrials on a rogue planet who can't imagine a planet near a star being inhabitable. An example of the second item in the list is seen in the *Star Trek* series, where spacefarers obey the Prime Directive, which forbids them from interfering with the development of other life forms. The last category is portrayed in the classic 1950s film *The Day the Earth Stood Still*, in which an extraterrestrial visitor warns that Earth will be destroyed unless we control our use of atomic weapons:

Klaatu barrada nikto!

All these schemes have two things in common. First, there is no evidence to support any of them, and, second, they are all somewhat improbable in a galaxy with thousands of different advanced civilizations. Some might indeed retreat to Dyson spheres or refuse to go near stars, but to suppose that all of them would is something of a stretch. Similar arguments can be made for the other explanations. We'll leave this subject, then, and go back to Occam's razor.

One way to approach the question posed by the Great Silence is to think of each term in the Drake equation as a gateway or valve on the way to an advanced technological civilization. If even one of those terms has a numerical value much less than we have assumed, the effect would be to greatly reduce our estimate of the number of extraterrestrials out there. In essence, that term would act as a kind of filter, blocking the orderly progression implied in the equation. To use a term introduced by economist Robin Hanson, our colleague at George Mason, somewhere in the chain of events in the Drake equation there might be a "Great Filter" that effectively blocks the development of civilizations that might be trying to communicate with us. We have already discussed many possible Great Filters, but let's review the terms in the Drake equation again to specify where they might be.

The first two terms in the equation, involving star and planet formation, are pretty solid. Thanks to the Kepler satellite, we know that planets are common, and astronomers have long known of regions in the Milky Way where stars are forming. The next two terms, however—the ones involving Earth-type planets and the development of life—are less certain. We've already discussed the role of the continuously habitable zone in diminishing the possibility of life, but there are other factors that could have the same effect. For example, if a planet circled a star much closer to the galactic center than the Sun, it would have many more nearby stellar neighbors. It is possible that in a crowded neighborhood, nearby supernovae or gamma ray bursts could wipe out life on planets near them. If this is the case, then planets not only have to be a certain distance from their star, but the star has to be a certain distance from the galactic center.

There are other things that might affect the development of life. For example, some theoreticians have suggested that the presence of a planet such as Jupiter plays an important role in sending water-rich asteroids into the inner solar system. It is this rain of asteroids, the argument goes, that produced Earth's oceans. Later, after the Late Heavy Bombardment, the Jupiter-like planet acts as a broom to eject large asteroids that cross its orbit and that might impact Earth and cause a mass extinction. So the Jupiter-like planet both produces the oceans and later shields the Earth-like planet. Although there is a good deal of debate on this subject, if it turns out that a Jupiter-sized planet is required to produce oceans and later be a shield, then the presence of an Earth-sized planet in a star's CHZ wouldn't be enough to produce life. You would need an appropriately located Jupiter as well.

If any of these effects really matter, then the Great Filter is right at the beginning of the process, and life will be very rare in the universe. No dinosaurs—not even any green pond scum—just a galaxy full of dead, sterile planets.

Moving to the right in the Drake equation, we have to start thinking about the development of intelligence. As we pointed out, this requires two separate evolutionary steps—the development of eukaryotes and the development of multicellularity. On Earth, each of these steps took about a billion years. If this is a typical time frame for these transitions, there has been ample time for both to occur on other planets. But what if the typical time frame for each is 10 billion years and we've just been lucky that they happened quickly here? What if the time frame is 100 billion years and we've just been *very* lucky? If this is the case, then the filter is located at the point of the development of intelligence and the galaxy will be full of green pond scum planets, but no complex life forms.

What about the development of a technological civilization? Here again there are large uncertainties, but it seems pretty obvious that it requires the evolution of high-level intelligence—something beyond the dinosaurs. On Earth, many scientists attribute the unusual development in the size of the human brain to events that took place in Africa 8 million years ago. This was a time when the forest that covered the eastern part of the continent was changing into a savannah, so that tree-dwelling primates who could walk from one patch of forest to the next had an evolutionary advantage. This led, the argument goes, to bipedal locomotion and the freeing of the hands for tool making and to the increased brain size to support it. Thus, if you accept this argument, you would have to say that in order to have extreme intelligence, the home planet must have the kind of plate tectonic activity that can produce this sort of environmental change. On the other hand, it's easy to imagine paths to high intelligence that don't involve plate tectonics—octopi are pretty smart, for example, as are dolphins, and neither requires special terrestrial conditions.

Having said this, we should note that large brains are not an unalloyed evolutionary advantage. In order for a human fetus to be able to fit through the female birth canal, human babies are born with their brains largely undeveloped. Thus, the evolutionary cost of the large brain is the requirement that infants must be supported by others for many years. It is hard to imagine what costs there might be to developing intelligence in an alien environment.

Further along in our evolutionary history, some scientists have argued that the existence of periodic ice ages played an important role in producing the kind of social interactions needed to take humans beyond the hunter-gatherer stage. In one scenario, for example, the need to protect the nutritionally rich shellfish beds along the African coast—a dependable source of food—during an ice age is what led to both the kind of cooperativeness and the kind of aggressiveness that have characterized our species ever since. Again, if you accept this sort of argument, you are saying that the Great Filter is located at the point where intelligence progresses into advanced society. If this is true, there will be lots of planets with the equivalent of dinosaurs out there, but none (or very few) with radio telescopes.

The kind of arguments we've been looking at—arguments that say, in effect, that there is something special about Earth that is unlikely to be duplicated elsewhere in the galaxy—go under the name of the Rare Earth Hypothesis. They are put forward most completely in a book titled *Rare Earth*, by geologist Peter Ward and astronomer Donald Brownlee. Ward and Brownlee's central thrust is that we have been blindly accepting the Copernican principle—the idea that Earth is not special—and ignoring the fact that there are many unusual things about our home planet. In essence, they look at all the things that are unique about Earth and argue that if they are all necessary for an advanced civilization to develop, then we could well be the only such civilization in the galaxy. For example, if, besides an Earth-sized planet in the CHZ of its star, you need a star located a certain distance from the galactic center, a Jupiter farther out, plate tectonics, the right planetary tilt to produce ice ages, and a large moon to stabilize the planet's axis of rotation and produce tidal pools (Darwin's warm little pond), Earth might well be the only planet like that in the galaxy. The Rare Earth answer to the Fermi paradox is thus quite simple: there's nobody here because there's nobody there. We are indeed alone.

Those who don't accept the Rare Earth Hypothesis assert that *any* specific event you want to talk about is extremely unlikely, and that simply reciting that fact proves nothing. Think, for example, of the chain of unlikely events that led to your reading these words. Your parents had to meet, you had to attend a certain school, learn to read, acquire an interest in science, and so on. There's no point in harping on this improbability, though, because if you weren't reading this book, you'd be doing something else equally improbable. In the same way, other types of improbable intelligences could have developed in the galaxy following their own improbable chain of events, and there could be an infinite number of those improbable paths. For these critics, all the Rare Earth Hypothesis proves is that there is at least one improbable path to an advanced civilization (our own); it says absolutely nothing about the possible existence of other paths.

Up to this point, we have examined various ways in which the Great Filter might have operated to limit the development of technological civilizations in the galaxy. The scenarios we have considered all have one thing in common: they all assume that the Great Filter is behind us, that by some combination of luck or providence, *Homo sapiens* has made it through all the filters and bottlenecks that stood in our way. But there is another, much more frightening possibility. What if none of these events in our past constitutes the Great Filter? What if the Great Filter is still in front of us?

Is There a Great Filter in Our Future?

To understand the importance of this question, let's think for a moment about the nature of the evolutionary process. Natural selection is driven by one criterion and one criterion only: the need to get an organism's genes into the next generation. Winners in the evolutionary game, in other words, are not determined by moral or ethical considerations. Consider the history of our own species as an example of this statement. The appearance of *Homo sapiens* in any region once we left Africa was accompanied by the disappearance of competing hominids (think Neanderthals and Denisovans) and just about every large animal (think woolly mammoths and giant tree sloths). We became the dominant life form on the planet by wiping out our competitors, either directly or indirectly. Given this history, we think it's fair to say that *Homo sapiens* is not the sort of species you'd want to meet in a dark alley, and the same will be true of any other winner of the evolutionary game who became the dominant species on their planet.

The "Great Filter is in front of us" argument goes like this: despite the Rare Earth Hypothesis, there really doesn't seem to be anything all that special about the way that life developed on Earth, and given the abundance of planets out there, there is no reason that complex life shouldn't be quite common. On the other hand, from what we know about the process of evolution, we can expect the winners of the evolutionary game on other planets to be no more benevolent than *Homo sapiens*. In this case, the coming Great Filter is easy to see. Once an aggressive, warlike species discovers science, they are likely to turn their discoveries against one another and, in essence, wipe themselves out.

The picture of galactic history that comes from this argument is a disturbing one. From the very beginning, intelligent, technologically advanced societies have appeared only to disappear in a short time as they succumb to their own dark inner nature—a nature produced by the laws of natural selection. No one is out there, in other words, because they've all wiped themselves out long ago, before we started listening.

This dour view of the role of intelligence and evolution, incidentally, is why so many prominent scientists have opposed what is called active SETI. This is the suggestion that instead of just listening for signals, we should send tight, powerful beams of radio waves toward stars likely to harbor intelligent life in their planetary systems. The reason for this objection is obvious. Any extraterrestrials we alert to our presence will be more advanced than we are—they could scarcely be less advanced, after all. Given the sad historical record in our own past of what happens when advanced and less-advanced groups encounter each other, the argument goes, our best strategy is to lie low and not call attention to ourselves. The authors see this as a sensible point of view.

Although the argument about the existence and possible location of a Great Filter seems abstract and philosophical, it is a question that can be answered by standard scientific means. If we find no life on Europa, no fossils on Mars, and no oxygen in the atmospheres of exoplanets, we can breathe a sigh of relief. Life is rare in the galaxy, and we have had the enormous luck to have gotten through whatever hurdles were in our way. If, however, we find evidence for green pond scum or numerous dinosaur planets, we would have to conclude that the Great Filter is in front of us. Given the current state of the world, neither of the authors finds that prospect particularly positive.

EPILOGUE

*The world is not only stranger
than you imagine, it's stranger
than you can imagine.*

Attributed to J. B. S. Haldane, Daedalus, or,
Science and the Future, 1923

S o, what have we learned in this voyage of exploration to
other worlds? In chapter 1, we identified three different
barriers to imagination—we called them "chauvinisms."
We can now see that two of these three—surface chauvinism and
stellar chauvinism—will have to be abandoned. The discovery of
subsurface oceans on the moons of the outer planets and Pluto
has shown that, insofar as water is necessary for the development
of life, it need not be found on the surface of a planet. In a similar
way, we shall argue below that the discovery of rogue planets and
the realization that they outnumber planets orbiting stars opens
exciting new vistas in the study of exoplanets.

Carbon chauvinism—the idea that life has to be based on molecules containing carbon chains—is a little harder to discount. Given the fact that we know of only one type of life—our own—there is simply no data that pushes us to give the idea up. We think that a strong case can be made that if life is based on chemistry, then that chemistry pretty much has to involve carbon.

It could be argued that this argument is just another example of chauvinism—call it chemical chauvinism. But whether it's because life really has to be based on chemistry or simply because of a failure of imagination, we have found it hard to imagine nonchemical life. Here, however, is one suggestion: it is possible to imagine a system of intertwined electric and magnetic fields inside a highly conducing metallic planet reaching a level of complexity comparable to that found in living systems. We don't know of any such systems, but that doesn't mean that they are impossible. We could even imagine that such systems could reproduce themselves and, of course, they would most likely need a source of energy. They would, in other words, have the sorts of characteristics we associate with life. We doubt, however, that we would recognize them as living systems. We have to keep an open mind on the issue of carbon chauvinism, but at the moment we see no reason to abandon it.

Having said this, however, we have to note that our current search strategies—concentrating on finding a Goldilocks planet and measuring atmospheric oxygen on exoplanets—are unlikely to turn up unexpected life forms. We are unlikely to find something that we aren't looking for, after all. But at the moment, no one (including the authors) knows how to go beyond the search for life that is "like us" in the sense of being based on chemistry.

Even if we eventually develop criteria for identification of non-chemical complexity at the level of life as we know it, we don't have tools as of yet for identifying such "entities" on distant planets.

From our point of view, however, the most exciting discovery in the search for exoplanets is the realization that most planets in the galaxy are not circling stars but are rogue planets. We visited one of these planets in chapter 7.

As we mentioned previously, estimates of the number of rogue worlds in the galaxy range from two to perhaps as many as 100,000 times the number of planets circling stars. Models suggest that during the period when it was forming, our own solar system kicked about 10 planet-sized objects out into space as the terrestrial planets formed, for example, and more from the outer planets. Given that the Sun formed late in the life of our galaxy—astronomers often refer to it as a "third-generation" star—it's not hard to see that the number of rogue planets would exceed the number that remained with their stars. The implication of this fact is that ever since the Green Bank conference (see chapter 13), we have been concentrating on a small fraction of planets in the galaxy.

This argument bears an uncanny resemblance to what has been happening in cosmology over the past couple of decades. Physicists spent most of the twentieth century trying to understand the structure of familiar matter. This stuff, composed of protons, neurons, and electrons, is what makes up our bodies and everything else we encounter on a daily basis. It's called baryonic matter. (*Baryon*, a term used to describe particles such as the proton and neutron, means "heavy one.") As the twentieth century wound down, however, we found that baryonic matter—the kind of stuff we had expended such effort to analyze—makes up only about 5 percent of the mass of the universe. The rest is made up of stuff called dark matter (23 percent) and dark energy (72 percent). In our quest to understand the basic structure of the universe, in other words, we have been concentrating on only a small fraction of what is out there. To us, the prevalence of rogue planets out there suggests that we have been doing the same thing with SETI.

Incidentally, despite the similarity in their names, dark matter and dark energy are very different things. To oversimplify, dark matter holds the stars in a galaxy together, while dark energy pushes the galaxies apart. The only thing they have in common is that we have no idea what either one is made of.

In any case, if you work out the numbers, you find that the distance between rogue planets can be anywhere from a little more than the distance between the Sun and the outer edge of our solar system to a little more than the distance to the nearest star—the answer depends on how many rogue planets there actually are. This result—especially the results giving smaller distances—has important consequences for SETI and for the arguments about galactic colonization we discussed in the last chapter.

The point is this: if there really are rogue planets floating around just outside the solar system, then the kind of multigenerational starships we discussed in chapter 14 won't be needed to reach them. We can easily imagine colonization missions taking less than 10 years, and making a rogue planet habitable probably wouldn't be any more difficult than doing the same thing on Mars.

But what if some of the rogue worlds have already produced life? How would scientists on those worlds see the universe?

First, they would see a galaxy in which worlds like their own were plentiful. They would almost certainly concentrate on looking at planets "like themselves." You can even imagine scientists on the rogue worlds arguing that life could not possibly exist on planets near stars, in a hostile environment full of ultraviolet radiation, solar storms, and asteroid impacts. Life could only exist, they might argue, in the tranquility of deep space. Why bother to try to communicate with worlds where life was clearly impossible? After all, we've spent most of our history ignoring rogue worlds. Why shouldn't they return the favor?

We argued in chapter 7 that life that developed on rogue worlds would not interact with its environment through the medium of visible light but through infrared radiation, or possibly radio waves. This means that when and if these life forms developed a science of astronomy, they would probably search for other sources of infrared radiation—other rogue worlds. Given the expected density of rogue worlds, communication between neighbors might be possible, and we might expect it to be done via tight infrared beams—beams that would be invisible to us unless we happened to stumble across one. These would, of course, represent a source of SETI outside of anything posited in the Drake equation.

So, the bottom line from our exploration of the world of exoplanets is that there is a lot more to be explored out there than we thought. Let's get on with the job!

index

butterfly analogy, exoplanets and, 2–3

Callisto (moon of Jupiter), 27, 131
Cambrian explosion, 171, 177
Cancer constellation, 80
carbon
    of exoplanets, 86–88
    extraterrestrial life and role of, 155–56
carbon-14 experiments, life on Mars and, 158
carbon chauvinism, 9–11, 156, 202
Cassini spacecraft, 24–25
Cech, Thomas, 150
Celestial Scenery, or the Wonders of the Planetary System Displayed, Illustrating the Perfections of Deity and a Plurality of Worlds (Dick), 16
cells, creation of, 146–48
    intelligence and, 175–79
center of mass, planetary orbits and, 43
centrifugal force, in planets, 56–57
Ceres, discovery and classification of, 50, 52, 56–57
Charon (moon of Pluto), 110–11
Chauvin, Nicolas, 10
chauvinisms, exoplanet research and, 10–12
chemical chauvinism, 202
chemical evolution, 144–46
Child, Julia, 146–47
CHNOPS mnemonic (carbon, hydrogen, nitrogen, oxygen, phosphorus, and sulfur), 162
Christianity, planetary research and conflicts with, 38–41

citric acid cycle, 152
Cocconi, Giuseppe, 169, 179–80
Coleridge, Samuel Taylor, 104
comets
    deuterium in, 131
    in Greek cosmology, 50
computerized data processing, planet validation and, 71–72
computer models of solar system, 32–33
Condemnations of Paris, 39
continuously habitable zone (CHZ), 120, 126, 164–65, 174, 182–84
cooling-off period
    energy for planets and, 93
    exoplanets and, 87–88
    for rogue planets, 98–99
Copernican principle, 171
Copernicus, Nicolaus, 16, 40, 50, 53
cosmology
    Aristotle's work in, 38–39
    current challenges in, 203–5
    Greek astronomers' view of, 36–37
    plurality of worlds debate and, 39–47
    solar system and, 15
Curiosity rover, 19, 158–59
curse of the single example, 2, 8, 20, 30
    Goldilocks planets and, 119
cyanobacteria, oxygen release by, 162–63
Cygnus constellation, 67–69

dark energy, 204–5
dark matter, 204–5
Darwin, Charles, 116, 141–42

Epicurus, 35
Eris (astral body), 29
eukaryotes, 175–79
  search for extraterrestrial
    intelligence and, 193–99
Europa (moon of Jupiter)
  Galilean measurements of, 3,
    21–23
  ice layer on, 22–23, 27
  possible life on, 23–24, 164
  water distribution on, 131, 156
  water plume on, 23–24
European colonialism, plurality of
  worlds debate and, 41
European Space Agency, 159–60
evolution
  cell creation and, 146–48
  chemical evolution, 144–46
  Drake equation and, 174–75
  Great Filter theory and, 196–99
  natural selection and, 141–44
  origin of life and, 124–26
  search for extraterrestrial life
    and, 194–99
ExoMars venture, 159
exomoons, possibility of, 164–66
exoplanets
  classification criteria for, 56–60
  definitions of, 60–61
  discovery of, 2–3
  early research on, 35–42, 42–43
  Kepler spacecraft identification
    of, 69, 74–76
  in Kuiper belt, 3–4
  life on, 8–10, 140–41, 152–53,
    160–64
  oceans on, 129–38
  rogue planets as, 203–5
  search for, 4–7

as steam and ice worlds, 133–34
twentieth-century research on,
  42–47
as water worlds, 132–38
exosociology, 172
extraterrestrial intelligence,
  168–84
extraterrestrial life
  early research on, 16–18
  on Mars, 156–60
  plurality of worlds debate and,
    40–47
  on rogue planets, 93
  search for, 155–66

false positives, in Kepler spacecraft
  identification, 70–72
fat globules, evolution and, 146
faux ideé claire (clear but false
  idea), early solar system
  research and, 17–18
Fermi, Enrico, 10, 186–87
Fermi Gamma-ray Space
  Telescope, 187
Fermi National Accelerator
  Laboratory, 187
Fermi paradox, 10, 186–99
55 *Cancri*, identification and
  characteristics of, 80–82
55 *Cancri* e, characteristics of,
  82–89
51 Pegasi, 54
  exoplanet around, 46
The First Men in the Moon (Wells),
  16–17
fossil record
  Cambrian explosion and, 171
  evolution and, 143–44
Francis (Pope), 41

FRESIP (Frequency of Earth-Sized Inner Planets), 73–74
frost line, 31
frozen accident theories, origin of life and, 124

Galactic Club, 173
galactic year, 187–90
Galápagos Islands, 141
Galilean moons
　of Jupiter, 20–22, 93
　for rogue planets, 98–100
Galilei, Galileo, 15–16, 21
Galileo mission, 20–24
Ganymede (moon of Jupiter), 27, 131
Genesis, Book of, 1:2, 128
George Mason University telescope, 70–71
Gilbert, W. S., 140
Gingerich, Owen, 55–56, 61
Gliese, Wilhelm, 133
Gliese 1214b, 132–34
Goldilocks planets, 7–8, 11, 20, 72, 111
　continuously habitable zone of, 120
　extraterrestrial life and, 155–56, 202–3
　Kepler 186f as example of, 119–26
Goldin, Daniel, 73–74
Grand Synthesis, 142
gravitational lensing, 95
gravitational pull
　exoplanet detection and, 5
　of Jupiter on Europa, 23
　on Kepler 186f, 122–23
　planetary formation and, 32–33

rogue planet detection and, 96–98
Great Filter, SETI searches and, 192–99
Great Silence problem, 181, 188–99
Greek astronomers
　planetary systems and research of, 36–38
　planets defined by, 50
Greek astronomy, planets in, 14–15
Green Bank Conference, 169–73
greenhouse effect
　for rogue planets, 100–101
　volcanic activity and, 134
Gretzky, Wayne, 168, 184

Haldane, J. B. S., 201
Hamlet (Shakespeare), 1
Hanson, Robin, 192
heat source, for Pluto, 111
heavy elements, star formation and production of, 80–81, 95
heliocentrism, solar system and, 16
helium "rain," 97, 109
Heller, René, 123
Herschel, William, 50–51
Hooker, J. D., 116
hot Earths, 7
hot Jupiters, 5, 74–75
House of Wisdom, 38
Hubble Space Telescope, 23, 42
　planet identification and, 70
human brain, size increase in, 194–95
Huygens, Christiaan, 26, 41
Huygens probe, 26

hydrogen
  star formation and role of, 117
  in water, 130–31
hydrostatic equilibrium, 57

ice ages, evolution and role of,
  195–99
ice line, 31
ice worlds, exoplanets as, 133–34
infrared radiation, for rogue
  planets, 100–101
intelligence, definitions of,
  175–79
intelligent design theory, 124
International Astronomical Union
  (IAU), classification of Pluto
  and, 54–60
International Bureau of Weights
  and Measures, 53
International Commission on
  Zoological Nomenclature, 54
International Union of Pure and
  Applied Chemistry (IUPAC),
  53
interstellar cloud
  deuterium formation and, 131
  star and planet formation from
    materials in, 80–81
Io (moon of Jupiter), 27

Jet Propulsion Laboratory, 29
John XXI (Pope), 38–39
Jovian planets, 31–33
  core of, 84
Jupiter
  asteroids from, 193
  center of mass and orbit of, 43
  in Copernican solar system,
    50–51

Galileo spacecraft voyage to,
  20–21
Galileo's study of, 15
gravitational effects on Europa
  of, 23
magnetic field of, 83–84
moons of, 3, 20–21, 93
tidal heating on, 93

Kepler, Johannes, 82–83
Kepler 186f, 119–26
  orbit of, 120–21
Kepler object of interest (KOI), 66,
  69–72
Kepler's laws of planetary motion,
  82–83
Kepler spacecraft, 5, 47, 60
  candidate planets identified by,
    69–72
  components and operation of,
    67–69
  development of, 72–74
  eclipsing binary contamination,
    69–72
  operating characteristics,
    66–69
  transit method of planet
    detection and, 63–66
Kuiper, Gerard, 3–4
Kuiper belt, 3–4
  early research on, 29
  formation of, 32
  on new planetary systems, 111,
    114
Kuiper belt objects (KBOs), 29–30,
  57
  energy for, 105–6
  Pluto as, 110–11, 114
  subsurface oceans on, 108

van de Kamp, Peter, 43–44
Venus
    in Copernican solar system,
        50–51
    early research on, 17–18
Viking spacecraft, Mars
    exploration of, 157–58

Ward, Peter, 196
water
    on Earth, 129–31
    on exoplanets, 131–38
    on Kepler 186f, 123
    on Mars, 19, 159–60
    molecular structure of, 135–36

on new planetary systems, 111
on outer planets, 21–28
phase change and, 108–9
on Pluto, 106–7
worlds of, 132–38
water ice
    on Europa (moon of Jupiter),
        22–23, 27
    on Pluto, 105–7
weak interaction in physics,
    187
Wells, H. G., 16–17
white dwarfs, 117
William of Ockham, 39, 190–91
Wolszczan, Aleksander, 44–46

photography
and illustration
credits